# SUMMIT MATH

*Learn at your **OWN** pace.*

# ALGEBRA 1

second edition

# 5

## FACTORING
## POLYNOMIALS &
## SOLVING
## QUADRATIC EQUATIONS

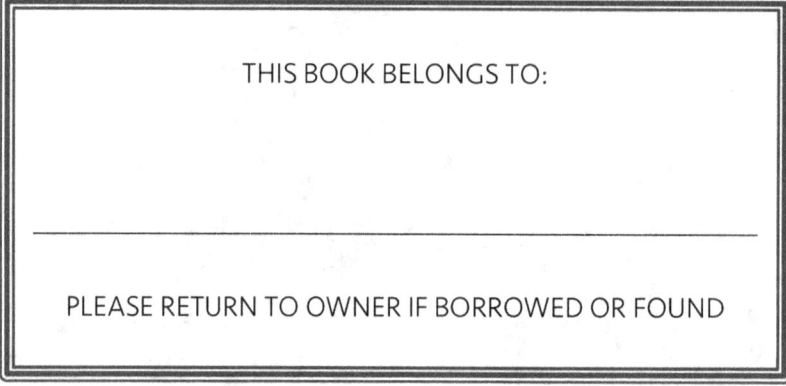

THIS BOOK BELONGS TO:

_____

PLEASE RETURN TO OWNER IF BORROWED OR FOUND

### DEDICATION
To Lauren, Chloe, Dawson and Teagan

### ACKNOWLEDGEMENTS
I started writing these books in 2013 to help my students learn better. I kept writing them because I received encouraging feedback from students, parents and teachers. Thank you to all who have used these books, pointed out my mistakes, and made suggestions along the way. Thank you to all of the students and parents who asked me to keep writing more books. Thank you to my family for supporting me through every step of this journey.

This book was typeset in the following fonts:
Seravek + Mohave + *Heading Pro*

Graphics in Summit Math books are made using the following resources:
Microsoft Excel | Microsoft Word | Desmos | Geogebra | Adobe Illustrator

First printed in 2017

Printed in the U.S.A.

Summit Math Books are written by Alex Joujan.

www.summitmathbooks.com

# INTRODUCTION

**Learning math through Guided Discovery:**
A Guided Discovery learning experience is designed to help you experience a feeling of discovery as you learn each new topic.

**Why this curriculum series is named Summit Math:**
Learning through Guided Discovery can be compared to climbing a mountain. Climbing and learning both require effort and persistence. In both activities, people naturally move at different paces, but they can reach the summit if they keep moving forward. Whether you race rapidly through these books or step slowly through each scenario, this curriculum is designed to keep advancing your learning until you reach the end of the book.

**Guided Discovery Scenarios:**
The Guided Discovery Scenarios in this book are written and arranged to show you that new math concepts are related to previous concepts you have already learned. Try to fully understand each scenario before moving on to the next one. To do this, try the scenario on your own first, check your answer when you finish, and then fix any mistakes, if needed. Making mistakes and struggling are essential parts of the learning process.

**Homework and Extra Practice Scenarios:**
After you complete the scenarios in each Guided Discovery section, you may think you know those topics well, but over time, you will forget what you have learned. Extra practice will help you develop better retention of each topic. Use the Homework and Extra Practice Scenarios to improve your understanding and to increase your ability to retain what you have learned.

**The Answer Key:**
The Answer Key is included to promote learning. When you finish a scenario, you can get immediate feedback. When the Answer Key is not enough to help you fully understand a scenario, you should try to get additional guidance from another student or a teacher.

**Star symbols:**
Scenarios marked with a star symbol ★ can be used to provide you with additional challenges. Star scenarios are like detours on a hiking trail. They take more time, but you may enjoy the experience. If you skip scenarios marked with a star, you will still learn the core concepts of the book.

**To learn more about Summit Math and to see more resources:**
Visit www.summitmathbooks.com.

As you complete scenarios in this part of the book, follow the steps below.

**Step 1: Try the scenario.**
Read through the scenario on your own or with other classmates. Examine the information carefully. Try to use what you already know to complete the scenario. Be willing to struggle.

**Step 2: Check the Answer Key.**
When you look at the Answer Key, it will help you see if you fully understand the math concepts involved in that scenario. It may teach you something new. It may show you that you need guidance from someone else.

**Step 3: Fix your mistakes, if needed.**
If there is something in the scenario that you do not fully understand, do something to help you understand it better. Go back through your work and try to find and fix your errors. Mistakes provide an opportunity to learn. If you need extra guidance, get help from another student or a teacher.

After Step 3, go to the next scenario and repeat this 3-step cycle.

Teaching videos for every scenario in the Guided Discovery section of this book are available at www.summitmathbooks.com/algebra-1-videos.

# CONTENTS

2

## Section 1
# REVIEW MULTIPLYING POLYNOMIALS

In later scenarios, you will eventually learn about a topic known as factoring. However, you first need to review how multiplying polynomials forms new polynomials.

1. Given the areas of the rectangles below, determine the values of $a$, $b$, $c$ and $d$.

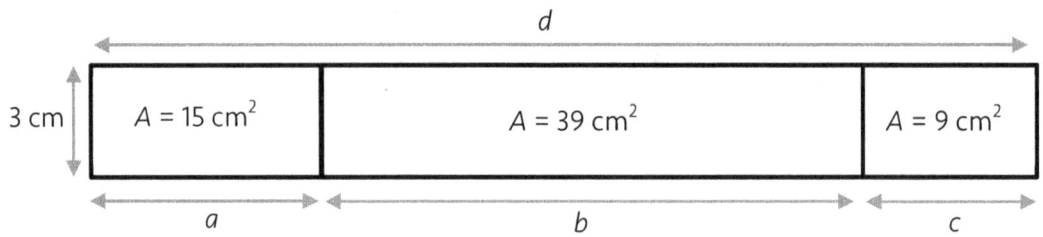

2. The area of the entire rectangle above can be expressed as 63 cm². If the area is written as the base times the height, it can be expressed as $(21 \cdot 3)$ cm². We'll be looking more closely at this later. Given the areas of the rectangles below, determine the lengths of $a$, $b$, and $c$.

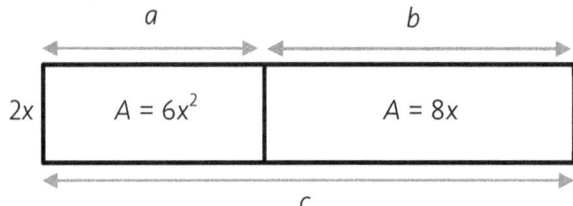

The area of the entire rectangle above can be expressed as $6x^2 + 8x$. If the area is written as the base times the height, it can be expressed as $(3x+4)(2x)$ or $2x(3x+4)$. The base is a <u>bi</u>nomial, 3x + 4, and the height is a <u>mono</u>mial, 2x, to review some vocabulary from earlier lessons.

3. Multiply each expression.

    a. $x(x+2)$          b. $3x(7x-5)$          c. $-4x(2x^2 - x - 6)$

4. Rewrite each expression as the product of a monomial and a binomial. How can you be sure that your product is a correctly rewritten form of the original expression?

    a. $5x^2 - 20x$          b. $6y^2 + 15y$

5. What are the values of *a*, *b*, and *c*, in the rectangle shown?  Two of the values are side lengths, and the other value is the area of the smaller rectangle.

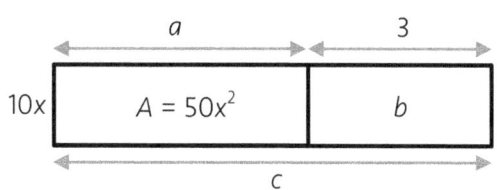

6. Draw a rectangle that has the following area.  Label the length of the base and height.

   a.  $11x+33$                         b.  $y^2+y$

7. Rewrite each of the areas in the previous scenario as the product of a monomial and a binomial.

   A brief lesson on vocabulary: when you rewrite an expression as the product of two other expressions, as in the previous scenario, the two expressions are called factors.  In this context, the word *factor* is a noun.  Additionally, the process of rewriting an expression as the product of its factors is often referred to as factoring.  In another context, then, the word *factor* is a verb.

8. Write each polynomial as the product of two factors.

   a.  $7x+7$                           b.  $10x^2-35x+5$

9. Factor each polynomial.

   a.  $3x^2-6x-5$                      b.  $8x^3+16x^2+12x$

   Identifying a common factor can lead to different opinions.  Consider the polynomial $8x^3+20x^2+12x$.  The three terms of this trinomial share 2 as a common factor, as well as 4 and 4x.  In situations like this, it is typical to identify 4x as the greatest common factor (GCF).

10. The terms in the expression below have many factors in common.  Identify all of the common factors of the terms in each expression.

   a.  $15x^3+30x^2+45x$                b.  $90x^3-30x^2+20x^4$

11. Identify the <u>greatest</u> common factor of each expression in the previous scenario.

12. Factor each polynomial below by writing it as the product of a monomial and a remaining polynomial. The monomial should be the greatest common factor of the original polynomial.

      a.  $2x^2 + 2x$           b.  $12x^3 + 30x^2 - 24x$          c.  $16x^4 - 80x^3 - 40x^2$

13. Factor the trinomial $18x^4 + 60x^3 - 30x^2$.

14. A rectangle is divided into 4 smaller rectangles as shown. Calculate the area of each of the smaller rectangles. Write the area of each smaller rectangle inside the space of that rectangle.

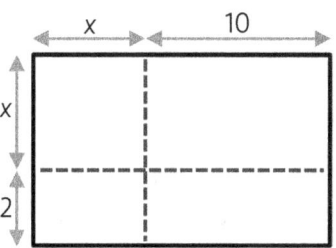

15. Consider the rectangle below. The areas of 3 of the sections are shown in those sections.

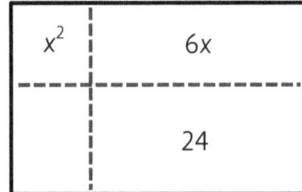

      a.  Determine the area of the missing section below.

      b.  Determine the area of the entire rectangle.

16. Consider the rectangle shown below.

      a.  Determine the expression for the area of the entire rectangle.

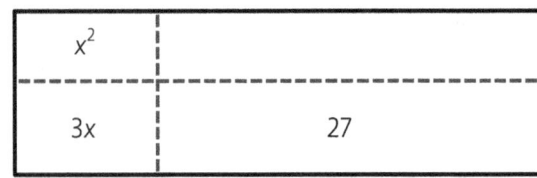

      b.  Write the area as the product of its base and height, or (base)(height).

# NOTES

Use this page to record important ideas in the previous section or for any other writing that helps you learn the topics in this book.

7

*Section 2*

# WRITING A TRINOMIAL AS A PRODUCT OF TWO BINOMIALS

17. In earlier lessons, you learned how to multiply binomials. Bring your mind back to that topic by multiplying the following binomials. Notice how the result relates to the original binomials.

 a. $(x+2)(x-7)$

 b. $(x-6)(x+6)$

 c. $(x-3)^2$

18. Look at your work for the 3 multiplication scenarios in the previous scenario. Could you do that work in reverse? If you started with your final expression, could you work backwards to figure out the two binomials that you multiplied to make that expression? Try to do that with the following expressions. Write each expression as the product of two binomials.

 a. $x^2-2x-3$

 b. $x^2-25$

 ★c. $x^2-8x+16$

19. What value of A makes the two expressions equivalent?

 a. $x^2+Ax-18$ and $(x+6)(x-3)$

 b. $x^2+Ax-121$ and $(x+11)(x-11)$

 c. $x^2+Axy+9y^2$ and $(x-3y)^2$

20. It may be helpful to go through the mental cycle of multiplying binomials again. As before, multiply the following binomials. Try to make connections between your final expression and the original binomials that you multiplied.

 a. $(x+3)(x-5)$

 b. $(x-7)(x+7)$

 c. $(x+2y)^2$

21. Once again, try to work backwards to write each expression as the product of two binomials.

 a. $x^2+7x+10$

 b. $x^2-y^2$

 ★c. $y^2-12y+36$

The single most important mental activity involved in factoring is the process of multiplying your factors together to check your answer.

22. Fill in the blank to make each expression equal to the factors written below the expression.

a.
$$x^2 + \underline{\phantom{xx}}x - 21$$
$$(x+7)(x-3)$$

b.
$$x^2 + \underline{\phantom{xx}}x - 6$$
$$(x-6)(x+1)$$

c.
$$x^2 - 14x + 24$$
$$(x + \underline{\phantom{xx}})(x-2)$$

23. Multiply the expressions below to find an equivalent expression without parentheses.

$$4(x+1)(x+3)$$

24. When you try to multiply the expressions above, it may be confusing to know how to proceed with the "4" to the left of the first binomial. You may find it easier to first multiply the binomials before distributing the "4", but you can also start by distributing the "4" to the first binomial. Try a third option. Distribute the "4" to both of the binomials and then multiply the binomials. Write down your simplified result to see if it matches the other two options.

25. Multiply the expressions below to find an equivalent expression without parentheses.

$$-2(x-5)(x+1)$$

26. Fill in the blank to make each expression equal to the factors written below the expression.

a.
$$2x^2 + \underline{\phantom{xx}}x - 4$$
$$2(x-2)(x+1)$$

b.
$$x^3 + \underline{\phantom{xx}}x^2 - 10x$$
$$x(x+2)(x-5)$$

c.
$$-x^2 - 3x + 18$$
$$-(x + \underline{\phantom{xx}})(x+6)$$

27. For the trinomial $2x^2 - 10x - 12$, which factorization is correct?

$$2(x-3)(x+2) \quad \text{or} \quad 2(x-3)(x-2)$$

28. Each trinomial below has been factored but the plus or minus signs are missing. Fill in either a plus sign (+) or a minus sign (−) to make the factors correct.

a.
$$x^2 + 7x + 10$$
$$(x \quad 2)(x \quad 5)$$

b.
$$x^2 + 8x + 7$$
$$(x \quad 7)(x \quad 1)$$

c.
$$x^2 + 35x + 124$$
$$(x \quad 4)(x \quad 31)$$

29. Each trinomial below has been factored but the signs are missing. Fill in the missing signs to make the factors correct.

a.
$$x^2 + 7x - 18$$
$$(x \quad 2)(x \quad 9)$$

b.
$$x^2 - 5x - 6$$
$$(x \quad 6)(x \quad 1)$$

c.
$$x^2 + 27x - 160$$
$$(x \quad 5)(x \quad 32)$$

30. Write each trinomial as the product of two binomials. This is also called <u>factoring</u> the trinomial.

a. $x^2 + 3x + 2$    b. $x^2 - 3x + 2$    c. $x^2 - x - 2$    d. $x^2 + x - 2$

31. Factor each polynomial below by writing it as the product of a monomial and a remaining polynomial. The monomial should be the greatest common factor of the original polynomial.

a. $4x^2 - 20x$    b. $-6x^2 + 18x - 3$    c. $14x^3 - 7x^2 - 35x$

32. Each trinomial below has been factored but some details are missing. Fill in any missing details to make the factors correct.

a.
$$x^2 - 13x + 12$$
$$(x \quad 12)(x \quad 1)$$

b.
$$x^2 - 11x + 24$$
$$(x \quad 3)(x \quad 8)$$

c.
$$x^2 - 55x + 600$$
$$(x \quad 40)(x \quad 15)$$

33. Factor each trinomial. Check your answer by multiplying your factors together. By doing this, you can prove that your factors are correct.

   a. $x^2 + 6x - 7$

   b. $100x^2 + 600x - 700$

The previous scenario is designed to show you that it helps to start out by *trying* to factor out a GCF from the entire expression. You may not be *able* to factor out a GCF, but if you can, it may help you find additional factors more easily.

34. Factor each trinomial below.

   a. $x^5 + 6x^4 - 7x^3$

   b. $-2x^2 - 12x + 14$

   ★c. $-3x^2 - 18x + 21$

35. Factor the following trinomials. Verify that your factors are correct by multiplying them together.

   a. $2x^2 + 4x - 6$

   b. $-3x^2 - 21x - 30$

36. Factor the trinomial $3x^2 + 4x - 7$.

Factoring the trinomial in the previous scenario can be challenging. It is easier to factor a trinomial when the leading coefficient is 1 (as in $1x^2$, or simply $x^2$). If the leading coefficient is any number other than 1, you need to pay attention to that coefficient and carefully check your factors.

37. Identify the leading coefficient in each trinomial. Rewrite each trinomial in Standard Form.

   a. $1 + 3x^2 - x$

   b. $2x + x^2 - 9$

   c. $3 - 8x - x^2$

38. Factor the following trinomials. Check your answer by multiplying your factors together.

   a. $11x + 5 + 2x^2$

   b. $5 + 16x + 3x^2$

   c. $4x^2 + 4x - 15$

39. What value of $B$ makes the two expressions equivalent?

    a. $3x^2 + Bx + 7$ and $(x-7)(3x-1)$          b. $10x^2 + Bx - 18$ and $2(5x+9)(x-1)$

40. Factor the following trinomials. Check your answer by multiplying your factors together.

    a. $7x - 6 + 3x^2$          b. $7 + 4x^2 - 16x$          c. $-5x^2 + 12 - 17x$

At this point, there isn't much more to learn about factoring. In order to strengthen your long-term understanding of this topic, though, we will continue to look more closely at factoring and try to sort out some of the patterns that may help you factor expressions more quickly.

41. Factor the following trinomials. Check your answer by multiplying your factors together. By doing this, you can prove that your factors are correct.

    a. $2x^2 - x - 6$                 b. $20x^2 - 10x - 60$

Once again, the previous scenario is designed to show you that it helps to start out by *trying* to factor out a GCF from the entire expression. You may not be *able* to factor out a GCF, but if you can, it may help you find additional factors more easily.

42. Factor each trinomial below.

    a. $2x^3 - x^2 - 6x$          b. $-6x^2 + 3x + 18$          c. $30 + 5x - 10x^2$

43. Explain why $(x-3)(x-2)$ cannot be the factorization of $x^2 - 5x - 6$.

44. Layla claims to know the factors for $2x^2 + 16x - 40$. Her factorization is written down as $(2x-4)(x+10)$. Is there another way to factor the original trinomial?

# NOTES

Use this page to record important ideas in the previous section or
for any other writing that helps you learn the topics in this book.

*Section 3*

# FACTORING A DIFFERENCE OF TWO SQUARES

45. Factor the following expressions.  Check your answer by multiplying your factors together.

   a.  $x^2+0x-4$

   b.  $x^2+0x-9$

   c.  $16+0x-x^2$

46. Factor the following expressions.  Check your answer by multiplying your factors together.

   a.  $25+0x-x^2$

   b.  $x^2-49$

   c.  $81-x^2$

47. Factor the following expressions.  Check your answer by multiplying your factors together.

   a.  $x^2-100$

   b.  $y^2-25$

   c.  $m^2-400$

48. Factor the following expressions.  Check your answer by multiplying your factors together.

   a.  $2x^2-8$

   b.  $2x^2-18$

   c.  $48-3x^2$

49. Factor the following expressions.  Check your answer by multiplying your factors together.

   a.  $75-3x^2$

   b.  $x^3-49x$

   c.  $-2x^2+72$

50. If $x^2-36$ can be factored as $(x+6)(x-6)$, how can $x^2+36$ be factored?

51. Factor the following expressions.

     a. $4x^2 - 49$                                         b. $16x^2 - 25y^2$

52. Factor the following expressions.  Check your answer by multiplying your factors together.

     a. $9f^2 - 64$                b. $64g^2 - 4$                c. $16h^2 + 25$

53. Factor the following expressions as much as you can.

     a. $x^4 - 81$                                          b. $x^4 - 1$

When a polynomial has a structure of $A^2 - B^2$, this structure is known as a difference of two squares. When it has this structure, its factors are the two binomials: $(A + B)(A - B)$.  Notice that in the previous scenario, the polynomials factored once and then the remaining expressions could be factored again.

54. Factor the expression $x^8 - 1$ as much as you can.

# NOTES

Use this page to record important ideas in the previous section or for any other writing that helps you learn the topics in this book.

Section 4
# FACTORING A PERFECT SQUARE TRINOMIAL

55. Factor the following trinomials.  Check your answer by multiplying your factors together.

     a.  $x^2+2x+1$               b.  $x^2-4x+4$            c.  $x^2-6x+9$

56. Factor the following trinomials.  Check your answer by multiplying your factors together.

     a.  $4x^2+20x+25$         b.  $1-10x+25x^2$        c.  $36-36x+9x^2$

57. What pattern do you notice in the previous two scenarios?

58. Factor the following trinomials.

     a.  $y^2-2xy+x^2$                    b.  $4x^2-28xy+49y^2$

59. Factor the following trinomials.

     a.  $-28x+2x^2+98$       b.  $-x^3-18x^2-81x$      c.  $-x^2+10xy-25y^2$

When you factor a trinomial and it has two factors that are exactly the same, it is called a <u>perfect square trinomial</u>.  Just as 16 is a perfect square because it can be written as $4^2$, a trinomial is a perfect square if it can be written as a binomial squared.

60. One type of perfect square trinomial will always factor into $(A+B)(A+B)$ or $(A+B)^2$.  Multiply these factors back together to see the original trinomial.

61. A perfect square trinomial may also sometimes factor into $(A-B)(A-B)$ or $(A-B)^2$.  Multiply these factors back together to see the original trinomial.

           20            

62. Factor the expression $A^2 - 20AB + 100B^2$.

63. Explain why $(x-15)(x-4)$ cannot be the factorization of $2x^2 - 19x + 60$.

64. Your classmate claims to have the factors for $x^3 - 7x^2 - 18x$. His factorization is written down as $(x^2 - 9x)(x+2)$. Is there another way to factor the original trinomial?

65. Consider the trinomial $25 - 10x + x^2$. If you factor it as $(5-x)^2$, but I factor it as $(x-5)^2$, how can we figure out whose factors are correct?

66. Factor the following trinomials.

    a. $A^2 + 2AB + B^2$         b. $A^2 - 2AB + B^2$         c. $-30x + 9x^2 + 25$

67. Explain how to determine if a trinomial is a perfect square trinomial.

68. Factor the following trinomials.

    a. $2x^2 - 9x - 11$         b. $3x^2 - 14x + 8$         c. $4x^2 + 12x - 7$

# NOTES

Use this page to record important ideas in the previous section or for any other writing that helps you learn the topics in this book.

## Section 5
# *USING FACTORING TO SOLVE EQUATIONS*

Let's take a break from factoring to consider some scenarios that contain polynomials like the ones we have been working with so far.

69. Suppose you attempt a basketball foul shot. The path of the ball as it travels through the air is modeled by the equation $H = -16t^2 + 20t + 6$, where $H$ is the height of the ball, measured in feet, $t$ seconds after it leaves your hand. What is the height of the ball at the exact moment the ball is released?

70. In the previous scenario, suppose the shot misses the basket completely and hits the ground below the basket. How long after the ball is released does it hit the ground? Try to set up an equation that would help you answer this question. If this is confusing, keep reading.

When the ball hits the ground, its height is 0. In the equation $H = -16t^2 + 20t + 6$, if you replace $H$ with 0, the equation becomes $0 = -16t^2 + 20t + 6$. You can try to isolate "$t$" in this equation, but you probably do not know how to solve this type of equation yet.

In the next scenarios, you will learn about equations that involve $w^2$, $t^2$, $n^2$, or even $x^2$. First, consider this: When you solve an equation like $2x - 6 = 17$, the variable appears in only one term. As a result, you can undo the operations to isolate the variable $x$. You can add 6 to undo the subtracted 6 and you can divide by 2 to undo the multiplied 2. Once you understand how to undo every operation, you can always isolate $x$ in this type of equation.

71. Now consider an equation like $x^2 + 5x = -6$.

    a. You can try to combine the two terms containing "$x$" but they are not **like** terms.

    b. If you undo the multiplied 5 in "$5x$" by dividing both sides by "5" you will end up with $\dfrac{x^2}{5} + x = \dfrac{-6}{5}$. This looks even more confusing.

    c. You can factor out an $x$ to get $x(x+5) = -6$ and then divide both sides by $(x + 5)$, but the equation becomes $x = \dfrac{-6}{x+5}$, which also looks very confusing.

You can keep trying to isolate the variable to solve the equation, but you need a new strategy.

72. Start by moving all of the terms to the left side of the equation to make it look like $x^2 + 5x + 6 = 0$. Now put your factoring practice to work and rewrite the trinomial as the product of two factors.

73. When the trinomial is factored, the equation becomes $(x+3)(x+2)=0$. As a reminder, the goal is to find all possible values of $x$ that make the equation true. Spend some time trying out different numbers until you find a number that makes the left side of the equation equal 0.

74. If it is difficult for you to find $x$-values that make the previous equation equal 0, it is easier if you think about each factor separately. The expression $(x+3)(x+2)$ has a value of "0" if $x+3=0$ or if $x+2=0$.

    a.  $x+3$ equals 0 if $x$ = _____.          b.  $x+2$ equals 0 if $x$ = _____.

75. In the previous scenario, $x$ can equal 2 values: –2 or –3. This means that the original equation, $x^2+5x=-6$, has 2 solutions. Check both of these solutions to confirm they are solutions.

    a.  Replace $x$ with –2.                     b.  Let $x$ = –3.

    $$(\quad)^2+5(\quad)=-6$$                    $$(\quad)^2+5(\quad)=-6$$

76. When you first learn how to solve equations, you typically see that an equation has 1 solution. In this book, you learn that some equations have 2 solutions (or more...). Looking at each equation below, how many solutions does the equation have? Do not solve each equation.

    a.  $x-7=0$          b.  $(x+5)(x-10)=0$          c.  $(x-5)(x+3)(x-2)=0$

77. When a polynomial is written in its factored form, it is easier to figure out how to make the polynomial equal 0, because you can look at each factor separately. Fill in the blanks below.

    a.  The expression $(x-3)(x-4)$ has a value of "0" if _____ $=0$ or if _____ $=0$.

    b.  The expression $(x+2)(x+5)$ is equal to "0" if _____ $=0$ or if _____ $=0$.

    c.  The equation $(3x-1)(2x+3)=0$ is true if _____ $=0$ or if _____ $=0$.

78. Solve this puzzle: There are two numbers. When you multiply the numbers together, their product is zero. One of the numbers is 7. What is the other number?

79. What value of x makes each equation true?

    a. $x-1=0$            b. $0=x+1$            c. $x-7=0$

80. What x-values make each equation true?

    a. $(x-2)(x+3)=0$       b. $(x+7)(x-1)=0$       c. $(x-4)(x-5)=0$

81. The equation $(A)(B)=0$ is true if $A = 0$. It is also true if $B = 0$. Suppose, instead that $(A)(B)=1$. What values of A and B make this statement true?

82. Solve the equation $(x-2)(x+3)=1$ by setting both factors equal to 1. Check your solutions to see if either one makes the original equation true.

83. Which equation is easier to solve? Why is this?

    Equation 1: $(x-1)(x-2)=12$        Equation 2: $(x-1)(x-2)=0$

84. What x-value makes each equation true?

    a. $(x+1)(x+3)=0$       b. $(x+6)(x+4)=0$       c. $(x-8)(x+2)=0$

85. Solve the following equations.

     a.  $x^2+4x+3=0$           b.  $x^2+10x+24=0$           c.  $x^2-6x-16=0$

86. Solve the equation $x^2-7x+10=0$.

87. Solve the equation $x^2-3x=70$.

88. Solve the equation $x^2+36=12x$.

89. Solve the equation $(x-4)(x-3)=2$.

90. What $x$-value makes the equation true?

     a.  $4x-1=0$                     b.  $5x+3=0$

91. What values of $x$ make each equation true?

      a. $(6x-1)(2x-1)=0$        b. $(3x+2)(3x-2)=0$        ★c. $(3x+4)(5x-1)=0$

92. What values of $x$ make each equation true?

      a. $(2x-3)(x-2)=0$        b. $(2x-5)(2x-5)=0$        ★c. $(3x+5)(x-1)=0$

93. Solve the following equations.

      a. $2x^2-7x+6=0$        b. $4x^2-20x+25=0$        ★c. $3x^2+2x-5=0$

94. Solve the equation $5x^2-9x-2=0$.

95. Solve the equation $25x^2+4=20x$.

96. Solve the equation $6x^2 + 11x = 10$.

97. The next group of equations may seem confusing because they have a slightly different structure than the previous ones that you have seen so far.  What values of $x$ make each equation true?

    a. $x(x-4) = 0$ 　　　　　 b. $x(x+7) = 0$ 　　　　　 ★c. $2x(4x-1) = 0$

98. In each of the previous three equations, one of the solutions is 0.  Why does this occur?

99. Solve the following equations.

    a. $x^2 - 8x = 0$ 　　　　　 b. $2x^2 + 6x = 0$ 　　　　　 ★c. $5x^3 - 30x^2 = 0$

100. What values of $x$ make each equation true?

    a. $2x^2(x+9) = 0$ 　　　　　 b. $4x^3(x-10) = 0$ 　　　　　 ★c. $5x(x^2-4) = 0$

101. ★Solve the following equations in two different ways.  For the first approach, factor as you have learned to do in the previous scenarios.  For the second approach, divide both sides by $x$ and then proceed with factoring.  What do you notice?

    a.  $x^2 = 3x$                                     b.  $x^3 - 8x^2 + 12x = 0$

102. Explain how you would start to solve the equation $x(x-4) = 5$.

103. Solve the following equations.

    a.  $2x^3 - 10x^2 + 12x = 0$                     b.  $4x^3 = 100x$

104. Solve the following equations.

    a.  $-6x^2 + 54x + 60 = 0$                       b.  $-9x^2 + 36 = 0$

105. ★Solve the following equation. You first saw this equation in a previous scenario, before you had learned how to use factoring to solve equations.

$$0 = -16t^2 + 20t + 6$$

106. Solve the equation $(x+3)(x+2) = 2$. Check your solutions.

107. Solve the equation $(2x+1)(4x-1) = 14$.

108. ★Solve the equation $2x(x-1) = 12$.

109. ★How many solutions does the equation have?

$$x^3 + x = 2x^2$$

# NOTES

Use this page to record important ideas in the previous section or for any other writing that helps you learn the topics in this book.

*Section 6*
# SCENARIOS THAT INVOLVE FACTORING

110. The area of the entire rectangle below is 48 square units.  What is the value of $x$?

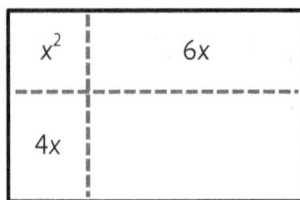

111. Although you missed the net on your earlier foul shot, you maintain your confidence and pick up the ball again, this time to attempt a half court shot.  On this shot attempt, the path of the ball is modeled by the equation $y = -16t^2 + 46t + 6$, where $y$ is the height of the ball, measured in feet, $t$ seconds after its release.

    a.  What is the height of the ball after 1 second?

    b.  Suppose the shot misses the basket completely again and hits the ground below the basket.  How long was the ball in the air before it hit the ground this time?

    ★c.  How many seconds after its release is the ball 25 feet above the ground?

112. The area of the rectangle shown is 24 cm$^2$.

    a.  What is the value of $x$?

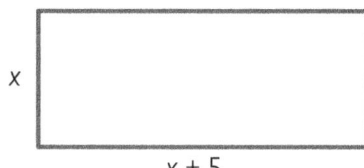

    b.  What is the perimeter of the rectangle?

113. ★If the area of the rectangle in the previous scenario is $9\frac{3}{4}$ cm², what are its dimensions?  This is easier to figure out if you can find a way to eliminate the fractions.

114. You want to build a rectangular frame with a length that is 8 inches more than its width.  If the side lengths must be integers and the area of the frame is 105 in², what are the dimensions of the frame?

115. The area of the triangle shown is 35 cm².  What is the length of the base of the triangle?

116. If the area of a triangle is 9 cm² and its height is 5 cm shorter than twice the length of its base, what are the dimensions of the triangle?

117. In the progression below, the 1st Figure contains 5 tiles and the 2nd Figure contains 10 tiles. Follow the pattern below and see if you can apply your algebraic fluency to this new scenario.

Figure 1          Figure 2          Figure 3

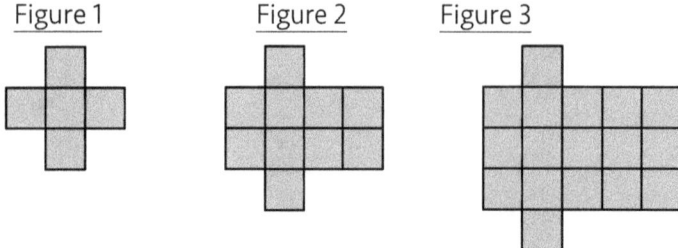

    a.  How many tiles will be in the 4th Figure?

    b.  How many tiles will be in the $n$th Figure?

    ★c.  Which Figure will contain 122 tiles? Answer this by creating and solving an equation.

118. Solve the following equations.

    a.  $3x - 5 = 4 + 2x$

    b.  $\dfrac{x-8}{3} = 5$

    c.  $\dfrac{x}{3} - 4 = 2$

119. Solve the following equations.

    a.  $x^2 - 8x = 0$

    b.  $2x^3 - 50x = 0$

    c.  $3x^2 - 5 = 2x$

There are many types of equations that you have learned how to solve (and more to come). Equations like $3x - 5 = 8$ and $-2x + 3 = x + 6$ usually have one solution and they are known as <u>linear</u> equations. Linear equations can be rearranged to look like $Ax + B = 0$. The equations that you have been learning about in recent scenarios look like $x^2 + 3x + 2 = 0$ and $x^2 + 5x = 6$. These equations are called <u>quadratic</u> equations and they can be rearranged to look like $Ax^2 + Bx + C = 0$.

120. Circle the equations that are quadratic.

    a.  $9x^2 - 2x + 3 = 0$         b.  $x^2 - 5 = 0$         c.  $x^2 = 9$         d.  $3x - 5 = 2x$

121. Solve the equation $x^2 = 9$.

122. ★The two solutions to a quadratic equation have a sum of 5 and a product of 6. The solutions are integers. What is the equation? Write the equation in the form $Ax^2 + Bx + C = 0$.

123. Determine the value of $x$ that makes the shaded region have an area of 46 square units.

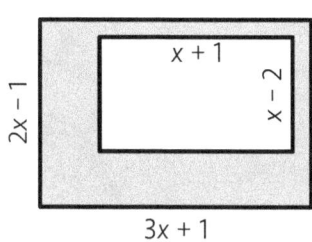

# NOTES

Use this page to record important ideas in the previous section or
for any other writing that helps you learn the topics in this book.

*Section 7*
# USING FACTORING TO SIMPLIFY FRACTIONS

124. Simplify each of the following fractions.

a. $\dfrac{1\cdot 5}{2\cdot 5}$

b. $\dfrac{6\cdot 2}{5\cdot 6}$

c. $\dfrac{2x}{5x}$

d. $\dfrac{10A}{40A}$

125. Simplify each of the following fractions.

a. $\dfrac{2(x-1)}{3(x-1)}$

b. $\dfrac{(x+5)(x+2)}{(x+3)(x+2)}$

c. $\dfrac{3x+6}{5x+10}$

126. In the previous scenario, the last fraction can only be simplified after you factor the numerator and denominator. After factoring, it becomes clear that there are identical factors in the numerator and denominator, which create a disguised form of 1. Use this strategy of factoring to simplify the following fraction.

$$\dfrac{x^2+3x+2}{x^2+6x+8}$$

127. Simplify each fraction by cancelling out disguised forms of 1.

a. $\dfrac{x^2-9}{x^2-2x-15}$

b. $\dfrac{3x^2-12}{x^2+7x-18}$

128. ★Simplify the following fraction: $\dfrac{8x^2-24x+18}{4x^2+10x-24}$.

# NOTES

Use this page to record important ideas in the previous section or for any other writing that helps you learn the topics in this book.

Section 8

# INTRODUCTION TO GRAPHING PARABOLAS

129. If you graph the equation $y = -x + 4$, what is the shape of your graph? If it is helpful, use the T-chart below to help you find points that are part of the graph. Plot at least 4 points.

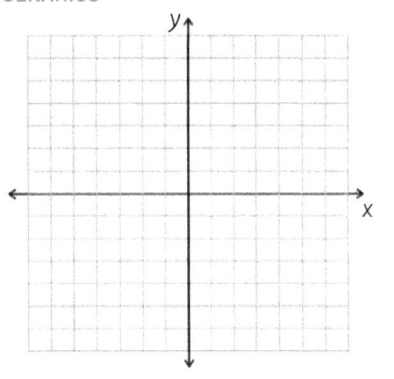

| x | y |
|---|---|
| 0 | $-(0) + 4 = 4$ |
| -1 | $-(-1) + 4 = 5$ |

130. When you graph all of the ordered pairs represented by the equation $y = x^2 - 4$, the shape of the graph is... actually, maybe you can figure out the shape of the graph. Use what you know about graphing equations to create a graph of this equation. A T-chart is started to help guide you.

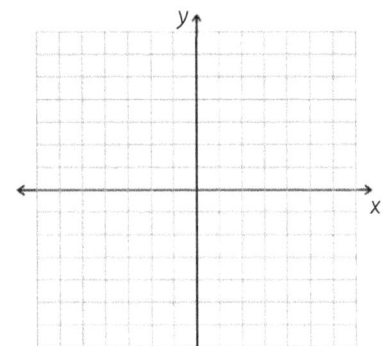

| x | y |
|---|---|
| 0 | $(0)^2 - 4 = -4$ |
| -1 | $(-1)^2 - 4 = -3$ |

The graph in the previous scenario has a curved shape that makes it very different than linear graphs. In mathematics, there is a technical name for practically everything. This specially curved graph is called a parabola (puh-RA-buh-luh). Additionally, while the structure $y = -x + 4$ is called a linear equation, the structure $y = x^2 - 4$ is called a quadratic equation.

131. Graph the equation $y = x^2 + 2x - 2$.

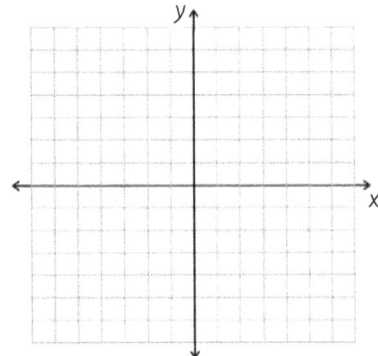

| x | y |
|---|---|
| 0 | $(0)^2 + 2(0) - 2 = -2$ |

132. Consider the parabola to the right.

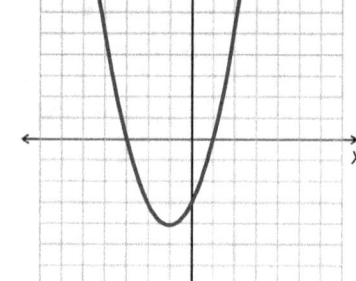

    a.  Find a point with a *y*-value of 0. What is the *x*-value?

    b.  Find a point with an *x*-value of 2. What is the *y*-value?

    ★c.  Which point on the parabola has a *y*-value of –5?

133. Determine the *x*-value(s) for which the equation $y = -2x^2 + 9x - 10$ has a *y*-value of

    a.  0                                      b.  –3

134. Identify any *x*- and *y*-intercepts for each of the following equations.

    a.  $4x - 7y = -14$                       b.  $y = -6x^2 + 7x + 5$

135. ★Identify any *x*- and *y*-intercepts for each of the following equations.

    a.  $x^2 + y = -9$                          b.  $y = 25 - 20x + 4x^2$

136. Identify the *x*- and *y*-intercepts of the following equation. Then, graph the equation to verify the accuracy of your intercepts.

$$y = x^2 - 4x$$

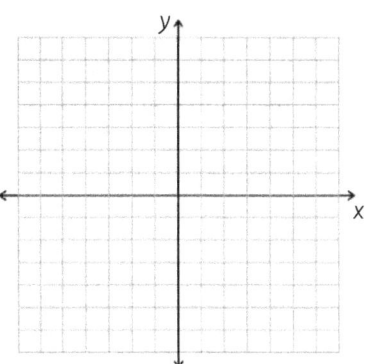

137. Draw a <u>very basic</u> sketch of a parabola that has each of the following characteristics.

a. <u>two *x*-intercepts</u>

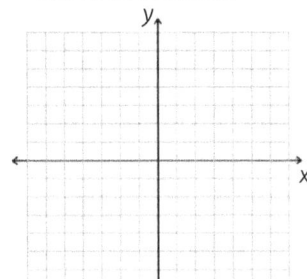

b. <u>one *x*-intercept</u>

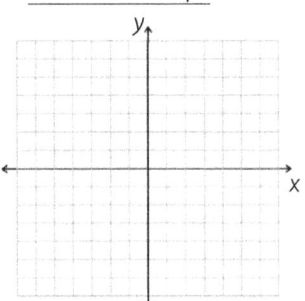

c. <u>no *x*-intercepts</u>

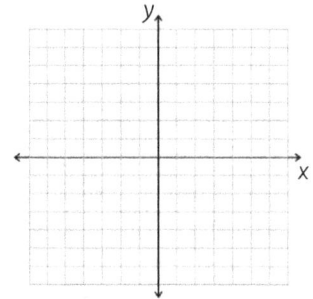

138. Part of a parabola is shown. Even though you do not have the equation for the parabola, use what you have learned so far to plot more points that are on the parabola. Use those points to finish drawing the parabola.

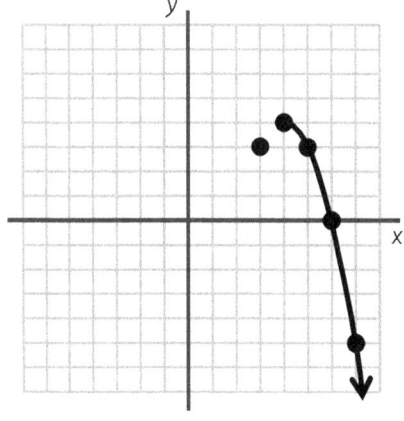

139. ★Using the grid, draw rectangles that have a perimeter of 24 inches. Draw as many different rectangles as you can discover. The only requirement is that each one must have a <u>perimeter</u> of 24 inches. Mark the length and width of each rectangle in the T-chart to the right of the grid.

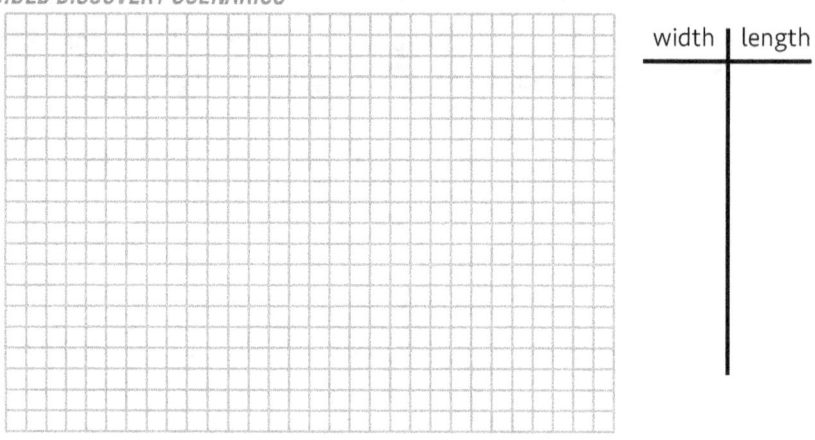

width | length

140. ★Calculate the area of each rectangle that you drew in the previous scenario. In the table shown, write the width and <u>area</u> measurements. Include more values in the table if you find more widths.

| Width | | | | | | | | | | | |
|-------|--|--|--|--|--|--|--|--|--|--|--|
| Area | | | | | | | | | | | |

141. ★In the graph shown, plot the data points from your table, with <u>width</u> on the horizontal axis and <u>area</u> on the vertical axis. Try to plot at least 10 points.

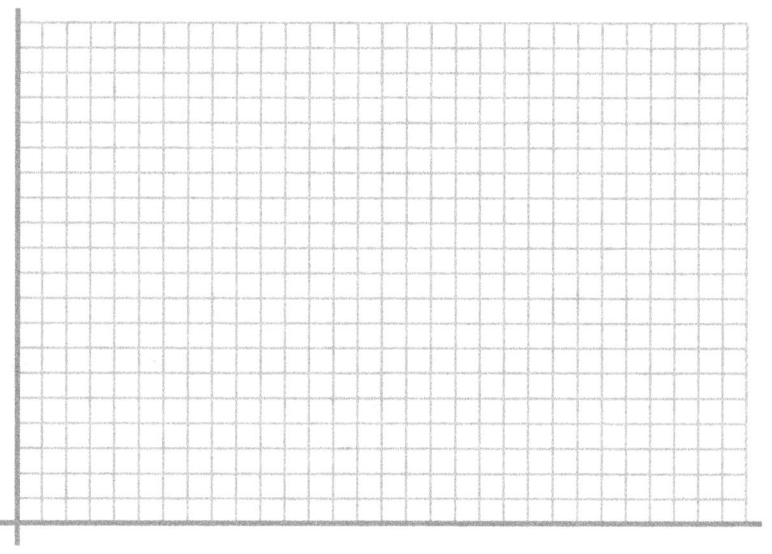

142. ★A rectangle has a perimeter of 24 inches.

    a. What is the length of the rectangle if its width is 4.5 inches? What is its area?

    b. What is the length of the rectangle if its width is $w$?

    c. What is the <u>area</u> of the rectangle if its width is $w$?

# NOTES

Use this page to record important ideas in the previous section or
for any other writing that helps you learn the topics in this book.

# Section 9
# CUMULATIVE REVIEW

143. Simplify each expression by eliminating parentheses and using only positive exponents.

    a. $\dfrac{10x^{-2}}{12x^{-5}}$
        b. $\left(\dfrac{-3x^{5}}{x^{-3}}\right)^{2}$
        c. $\left(\dfrac{2^{-1}a^{5}}{b^{-2}}\right)^{-3}$

144. It is often difficult to remember the difference between the area and perimeter of a rectangle. You have worked through several scenarios that involve the area of rectangles. Determine the perimeter of each rectangle shown below.

    a.
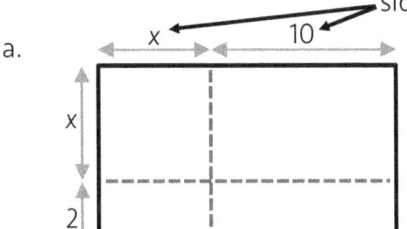

    b.

145. When drivers use the PA Turnpike, they pay a flat fee of $0.25 plus an additional 9.5 cents per mile. How many miles can a driver drive on the turnpike if she wants her toll to be at most $10, rounded to the nearest whole number?

146. A school ordered a new copy machine last year. The bill for the copy machine, which included 7% sales tax, was $868.84. How much did the copy machine cost before taxes were added to the price?

147. Solve each equation.

    a. $-2x + 3 = 3x - 7$                 b. $\frac{1}{2}x - 5 = -4x + 1$

148. What single operation could you perform to solve the equation in 1 step?

    a. $\frac{3}{5}x = 9$                 b. $-\frac{2}{3}x = 10$

149. Determine the equation of each line shown.

    a. 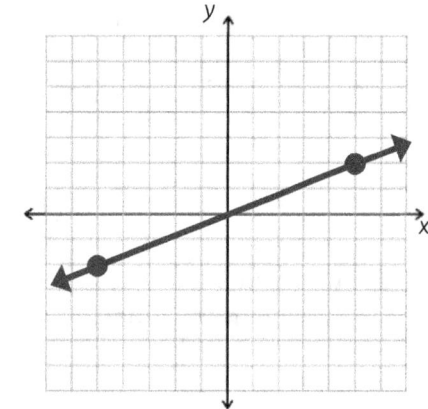      b.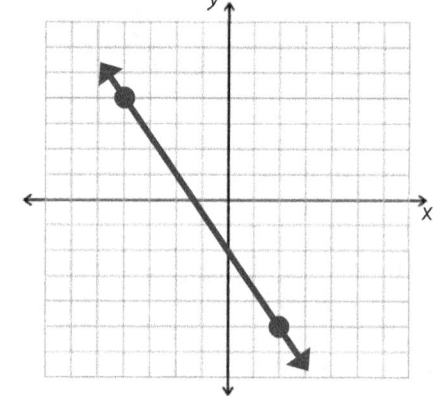

150. Determine the equation of each line shown.

    a. 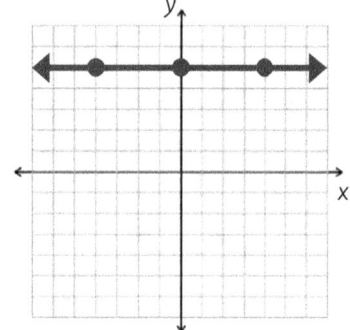      b.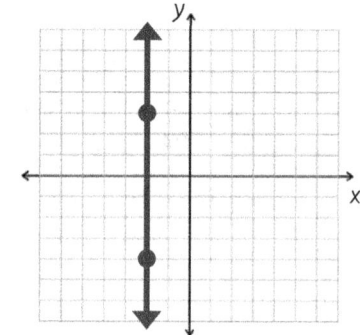

151. In the previous scenario, if you graph the two lines on the same grid, at what ordered pair would the two lines intersect?

# NOTES

Use this page to record important ideas in the previous section or
for any other writing that helps you learn the topics in this book.

## Section 10
# ANSWER KEY

| | |
|---|---|
| 1. | a. 5 cm    b. 13 cm    c. 3 cm    d. 21 cm |
| 2. | a. $3x$        b. 4        c. $(3x + 4)$ |
| 3. | a. $x^2+2x$            b. $21x^2-15x$ <br> c. $-8x^3+4x^2+24x$ |
| 4. | a. $5x(x-4)$        b. $3y(2y+5)$ |
| 5. | $a = 5x, b = 30x, c = 5x + 3$ |
| 6. | 11    $x+3$        $y$    $y+1$ |
| 7. | a. $11(x+3)$            b. $y(y+1)$ |
| 8. | a. $7(x+1)$            b. $5(2x^2-7x+1)$ |
| 9. | a. Prime (not factorable) <br> b. $4x(2x^2+4x+3)$ |
| 10. | a. $3, 5, 15, x, 3x, 5x, 15x$ <br> b. $2,5,10, x,2x,5x,10x, x^2,2x^2,5x^2,10x^2$ |
| 11. | a. $15x$            b. $10x^2$ |
| 12. | a. $2x(x+1)$        b. $6x(2x^2+5x-4)$ <br> c. $8x^2(2x^2-10x-5)$ |
| 13. | $6x^2(3x^2+10x-5)$ |
| 14. | $x^2$    $10x$ <br> $2x$    $20$ |
| 15. | a. $4x$    b. $x^2 + 10x + 24$ |
| 16. | a. $x^2 + 12x + 27$    b. $(x+9)(x+3)$ |
| 17. | a. $x^2-5x-14$    b. $x^2-36$    c. $x^2-6x+9$ |
| 18. | a. $(x+1)(x-3)$        b. $(x+5)(x-5)$ <br> c. $(x-4)^2$ |
| 19. | a. 3        b. 0        c. $-6$ |
| 20. | a. $x^2-2x-15$        b. $x^2-49$ <br> c. $x^2+4xy+4y^2$ |
| 21. | a. $(x+2)(x+5)$            b. $(x+y)(x-y)$ <br> c. $(y-6)^2$ |

| | |
|---|---|
| 22. | a. 4        b. $-5$        c. $-12$ |
| 23. | $4(x^2+4x+3)\rightarrow 4x^2+16x+12$ |
| 24. | When you distribute the 4 to both binomials, the result is 4 times larger than it should be. |
| 25. | $-2(x^2-4x-5)\rightarrow -2x^2+8x+10$ |
| 26. | a. $-2$    b. $-3$    c. $-3$ |
| 27. | Neither; $2(x-6)(x+1)$ is correct. |
| 28. | a. $(x+2)(x+5)$        b. $(x+7)(x+1)$ <br> c. $(x+4)(x+31)$ |
| 29. | a. $(x-2)(x+9)$        b. $(x-6)(x+1)$ <br> c. $(x-5)(x+32)$ |
| 30. | a. $(x+2)(x+1)$        b. $(x-2)(x-1)$ <br> c. $(x-2)(x+1)$        d. $(x+2)(x-1)$ |
| 31. | a. $4x(x-5)$        b. $-3(2x^2-6x+1)$ <br> c. $7x(2x^2-x-5)$ |
| 32. | a. $(x-12)(x-1)$        b. $(x-3)(x-8)$ <br> c. $(x-40)(x-15)$ |
| 33. | a. $(x+7)(x-1)$        b. $100(x+7)(x-1)$ |
| 34. | a. $x^3(x+7)(x-1)$        b. $-2(x+7)(x-1)$ <br> c. $-3(x+7)(x-1)$ |
| 35. | a. $2(x+3)(x-1)$        b. $-3(x+5)(x+2)$ |
| 36. | $(3x+7)(x-1)$ |
| 37. | a. $3; 3x^2-x+1$        b. $1; x^2+2x-9$ <br> c. $-1; -x^2-8x+3$ |
| 38. | a. $(2x+1)(x+5)$        b. $(x+5)(3x+1)$ <br> c. $(2x+5)(2x-3)$ |
| 39. | a. $-22$                b. 8 |
| 40. | a. $(3x-2)(x+3)$    b. $(2x-7)(2x-1)$ <br> c. $-(5x-3)(x+4)$  or  $(-5x+3)(x+4)$ |
| 41. | a. $(2x+3)(x-2)$    b. $10(2x+3)(x-2)$ |

52

| | |
|---|---|
| 42. | a. $x(2x+3)(x-2)$    b. $-3(2x+3)(x-2)$<br>c. $-5(2x+3)(x-2)$ |
| 43. | The last term would need to be "+6". |
| 44. | If you factor out the GCF first, you end up<br>with $2(x-2)(x+10)$. |
| 45. | a. $(x+2)(x-2)$    b. $(x+3)(x-3)$<br>c. $(4+x)(4-x)$ |
| 46. | a. $(5+x)(5-x)$    b. $(x+7)(x-7)$<br>c. $(9+x)(9-x)$ |
| 47. | a. $(x+10)(x-10)$    b. $(y+5)(y-5)$<br>c. $(m+20)(m-20)$ |
| 48. | a. $2(x+2)(x-2)$    b. $2(x+3)(x-3)$<br>c. $3(4+x)(4-x)$ |
| 49. | a. $3(5+x)(5-x)$    b. $x(x+7)(x-7)$<br>c. $-2(x-6)(x+6)$ |
| 50. | It cannot be factored. It is prime. |
| 51. | a. $(2x+7)(2x-7)$    b. $(4x+5y)(4x-5y)$ |
| 52. | a. $(3f+8)(3f-8)$    b. $4(4g+1)(4g-1)$<br>c. It cannot be factored. It is prime. |
| 53. | a. $(x^2+9)(x+3)(x-3)$<br>b. $(x^2+1)(x+1)(x-1)$ |
| 54. | $(x^4+1)(x^2+1)(x+1)(x-1)$ |
| 55. | a. $(x+1)^2$    b. $(x-2)^2$    c. $(x-3)^2$ |
| 56. | a. $(2x+5)^2$    b. $(1-5x)^2$ or $(5x-1)^2$<br>c. $9(2-x)^2$ or $9(x-2)^2$ |
| 57. | The factored form of each trinomial<br>contains a binomial that is squared. |
| 58. | a. $(y-x)^2$    b. $(2x-7y)^2$ |
| 59. | a. $2(x-7)^2$  b. $-x(x+9)^2$  c. $-(x-5y)^2$ |
| 60. | $A^2 + 2AB + B^2$ |
| 61. | $A^2 - 2AB + B^2$ |
| 62. | $(A-10B)^2$ |
| 63. | $x \cdot x$ is not $2x^2$ |
| 64. | If you factor out the GCF first, it becomes<br>$x(x-9)(x+2)$. |
| 65. | Both are correct |
| 66. | a. $(A+B)^2$    b. $(A-B)^2$    c. $(3x-5)^2$ |
| 67. | In a perfect square trinomial, the middle |

| | |
|---|---|
| | term is twice the product of the square<br>roots of the first and third terms. Stated<br>another way, find the product of the first<br>and third terms. Take the square root of<br>that result. If the middle term is twice<br>the value of that square root, the<br>trinomial is a perfect square trinomial. |
| 68. | a. $(2x-11)(x+1)$    b. $(3x-2)(x-4)$<br>c. $(2x-1)(2x+7)$ |
| 69. | 6 ft; In the equation, replace $t$ with 0. |
| 70. | You could solve the equation<br>$0=-16t^2+20t+6$, but you do not know<br>how to do this yet. This is intentional. In<br>the coming scenarios, you will learn how<br>to solve this type of equation. |
| 71. | – |
| 72. | $(x+3)(x+2)$ |
| 73. | $x = -3$ or $x = -2$ |
| 74. | a. $x = -3$    b. $x = -2$ |
| 75. | a. $(-2)^2+5(-2)=-6 \rightarrow 4-10=-6 \rightarrow$ TRUE<br>b. $(-3)^2+5(-3)=-6 \rightarrow 9-15=-6 \rightarrow$ TRUE |
| 76. | a. 1 solution    b. 2 factors: 2 solutions<br>c. 3 factors: 3 solutions |
| 77. | a. $x-3=0; x-4=0$<br>b. $x+2=0; x+5=0$<br>c. $3x-1=0; 2x+3=0$ |
| 78. | 0 |
| 79. | a. 1    b. –1    c. 7 |
| 80. | Each equation has 2 solutions.<br>a. x = 2 or –3  b. x = –7 or 1  c. x = 4 or 5 |
| 81. | There are infinitely many options.<br>$A=1, B=1$ or $A=2, B=\frac{1}{2}$ or $A=3, B=\frac{1}{3}$<br>or $A=4, B=\frac{1}{4}$ etc... |
| 82. | x = 3 or –2 but these solutions do not<br>make the original equation true. |
| 83. | #2; The product equals "0," which forces<br>one of the factors to be "0." |
| 84. | a. x = –3 or –1    b. x = –6 or –4<br>c. x = 8 or –2 |
| 85. | Factor each trinomial. When factored,<br>the equations in this scenario become the<br>equations in the previous scenario.<br>a. x = –3 or –1    b. x = –6 or –4<br>c. x = 8 or –2 |
| 86. | $(x-5)(x-2)=0 \rightarrow$ x = 5 or 2 |
| 87. | Rewrite the equation to make it equal 0. |

| | | | |
|---|---|---|---|
| | $x^2-3x-70=0\rightarrow(x-10)(x+7)=0$ <br> $\rightarrow x=10$ or -7 | 106. | $x^2+5x+6=2\rightarrow x^2+5x+4=0$ <br> $(x+4)(x+1)=0\rightarrow x=-1$ or -4 |
| 88. | $x^2-12x+36=0\rightarrow x=6$ | 107. | $8x^2+2x-1=14\rightarrow 8x^2+2x-15=0$ <br> $(4x-5)(2x+3)=0\rightarrow x=\frac{5}{4},-\frac{3}{2}$ |
| 89. | $x^2-7x+12=2\rightarrow x^2-7x+10=0$ <br> $\rightarrow x=5$ or 2 | | |
| 90. | a. $\frac{1}{4}$      b. $-\frac{3}{5}$ | 108. | $2x^2-2x=12\rightarrow 2x^2-2x-12=0$ <br> $2(x-3)(x+2)\rightarrow x=3$ or $-2$ |
| 91. | a. $x=\frac{1}{6},\frac{1}{2}$   b. $x=-\frac{2}{3},\frac{2}{3}$   c. $x=-\frac{4}{3},\frac{1}{5}$ | 109. | Two solutions. <br> $x^3-2x^2+x=0\rightarrow x(x^2-2x+1)=0$ <br> $\rightarrow x(x-1)(x-1)=0\rightarrow x=0$ or $x=1$ |
| 92. | a. $x=\frac{3}{2},2$   b. $x=\frac{5}{2}$   c. $x=1,-\frac{5}{3}$ | | |
| 93. | Factor each trinomial. When factored, the equations in this scenario become the equations in the previous scenario. <br><br> a. $x=\frac{3}{2},2$   b. $x=\frac{5}{2}$   c. $x=1,-\frac{5}{3}$ | 110. | $x^2+10x+24=48\rightarrow x^2+10x-24=0$ <br> $(x+12)(x-2)=0\rightarrow x=-12$ or 2 <br><br> A rectangle cannot have negative side lengths. The only $x$-value that makes the sides have positive lengths is $x=2$. Even though the equation has 2 solutions, one of the solutions can be ignored. |
| 94. | $(5x+1)(x-2)=0\rightarrow x=-\frac{1}{5}$ or 2 | | |
| 95. | $25x^2-20x+4=0$ <br> $(5x-2)(5x-2)=0\rightarrow x=\frac{2}{5}$ | 111. | a. 36 feet    b. 3 seconds <br> c. $\frac{1}{2}$ and $2\frac{3}{8}$ seconds |
| 96. | $6x^2+11x-10=0$ <br> $(3x-2)(2x+5)=0\rightarrow x=\frac{2}{3}$ or $-\frac{5}{2}$ | 112. | a. $x=3$    b. 22 cm |
| 97. | a. $x=0,4$   b. $x=0,-7$   c. $x=0,\frac{1}{4}$ | 113. | 1.5 x 6.5 <br> solve: $x(x+5)=\frac{39}{4}\rightarrow x^2+5x=\frac{39}{4}\rightarrow$ <br> $4x^2+20x-39=0\rightarrow(2x-3)(2x+13)=0$ <br> $x=1.5$ or $-6.5$ |
| 98. | The factor multiplied by the expression in parentheses can be replaced with 0 to make the equation true. | | |
| 99. | a. $x=0,8$   b. $x=0,-3$   c. $x=0,6$ | 114. | 7in x 15in    solve $w(w+8)=105$ <br> $w=7$ or $-15$ |
| 100. | a. $x=0,-9$   b. $x=0,10$   c. $x=0,2,-2$ | 115. | base: 10 cm <br> solve $\frac{1}{2}(x+5)(x+2)=35$ ; $x=5$ |
| 101. | a. $x=0,3$ vs. $x=3$ <br> b. $x=0,6,2$ vs. $x=6,2$ <br> When you divide both sides by $x$, you "lose" the ability to see that $x=0$ is one of the solutions. | | |
| 102. | Distribute the "$x$" first and then make the equation equal 0. <br> $x^2-4x=5\rightarrow x^2-4x-5=0$ | 116. | base = 4.5cm; height = 4cm <br> $\frac{1}{2}(x)(2x-5)=9\rightarrow 2x^2-5x=18$ <br> $2x^2-5x-18=0\rightarrow(2x-9)(x+2)$ <br> $x=4.5$ or $-2$ |
| 103. | a. $x=0,2,3$    b. $x=0,5,-5$ | | |
| 104. | a. $-6(x^2-9x-10)=0$ <br> $-6(x-10)(x+1)=0\rightarrow x=10,-1$ <br> b. $-9(x^2-4)=0\rightarrow -9(x+2)(x-2)=0$ <br> $x=2,-2$ | 117. | a. 26   b. $(n+1)^2+1$ or <br> $n(n+2)+2$ or $n^2+2n+2$ <br> c. 10th Figure; solve $n^2+2n+2=122$ |
| | | 118. | a. $x=9$   b. $x=23$   c. $x=18$ |
| | | 119. | a. $x=0,8$   b. $x=0,5,-5$ <br> c. $x=\frac{5}{3},-1$ |
| 105. | a. $0=-2(8t^2-10t-3)$ <br> $0=-2(4t+1)(2t-3)\rightarrow t=-0.25$ or 1.5 | 120. | Circle the equation in part a, b, and c. |
| | | 121. | 2 solutions: $x=3$ or $-3$ |

| | |
|---|---|
| 122. | $x^2 - 5x + 6 = 0$<br>Find pairs of integers with a product of 6:<br>1 and 6     2 and 3<br>The pair with a sum of 5 is 2 and 3.<br>Work backwards. If the solution is 2 and 3, the factors are $x - 2$ and $x - 3$.<br>$(x-2)(x-3) = 0 \to x^2 - 5x + 6 = 0$ |
| 123. | $x = 2$<br>solve: $(2x-1)(3x+1) - (x-2)(x+1) = 21$ |
| 124. | a. $\dfrac{1}{2}$   b. $\dfrac{2}{5}$   b. $\dfrac{2}{5}$   b. $\dfrac{1}{4}$ |
| 125. | a. $\dfrac{2}{3}$   b. $\dfrac{x+5}{x+3}$   c. $\dfrac{3(x+2)}{5(x+2)} \to \dfrac{3}{5}$ |
| 126. | $\dfrac{(x+2)(x+1)}{(x+2)(x+4)} \to \dfrac{x+1}{x+4}$ |
| 127. | a. $\dfrac{(x+3)(x-3)}{(x+3)(x-5)} \to \dfrac{x-3}{x-5}$   b. $\dfrac{3(x^2-4)}{(x+9)(x-2)}$<br><br>$\to \dfrac{3(x+2)(x-2)}{(x+9)(x-2)} \to \dfrac{3(x+2)}{x+9}$ |
| 128. | $\dfrac{2(4x^2-12x+9)}{2(2x^2+5x-12)} \to \dfrac{(2x-3)(2x-3)}{(2x-3)(x+4)} \to \dfrac{2x-3}{x+4}$ |
| 129. | |
| 130. | |
| 131. | |
| 132. | a. 2 $x$-values: 1 and –3   b. $y = 5$ |

| | |
|---|---|
| 133. | c. No point has a $y$-value of –5<br>a. $x = \dfrac{5}{2}, 2$   b. $x = \dfrac{7}{2}, 1$ |
| 134. | a. $(0, 2), (-3.5, 0)$<br>b. $(0, 5), \left(-\dfrac{1}{2}, 0\right), \left(\dfrac{5}{3}, 0\right)$ |
| 135. | a. $(0, -9)$, no $x$-intercept<br>b. $(0, 25), (2.5, 0)$ |
| 136. | <br>$x$-intercepts: (0,0), (4,0); $y$-intercept: (0,0) |
| 137. | Draw a parabola that. . .<br>a. crosses the $x$-axis 2 times<br>b. crosses the $x$-axis 1 time<br>c. does not cross the $x$-axis |
| 138. | Plot more points at (2,0) and (1,–5)<br> |
| 139. | width    length<br>1       11<br>2       10<br>3       9<br>4       8<br>5       7<br>6       6<br>7       5<br>8       4<br>...       ... |
| 140. | | Width | 1 | 2 | 3 | 4 | 5 | 6 | 7 | ... |<br>|---|---|---|---|---|---|---|---|---|<br>| Area | 11 | 20 | 27 | 32 | 35 | 36 | 35 | ... | |
| 141. | |
| 142. | a. $l = 7.5$ inches;  $A = 33.75$ in$^2$<br>b. length: $12 - w$   c. Area: $w(12 - w)$ |

| | |
|---|---|
| 143. | a. $\dfrac{5x^3}{6}$  b. $9x^{16}$  c. $\dfrac{8}{a^{15}b^6}$ |
| 144. | a. $4x + 24$  b. $4x + 20$ |
| 145. | ≤102 miles; solve $0.25 + .095m \le 10$ |
| 146. | $812.00; solve $x + .0.07x = 868.84$ |
| 147. | a. $2 = x$  b. $x = \dfrac{4}{3}$ |

| | |
|---|---|
| 148. | a. Multiply both sides by $\dfrac{5}{3}$<br><br>b. Multiply both sides by $-\dfrac{3}{2}$ |
| 149. | a. $y = \dfrac{2}{5}x$  b. $y = -\dfrac{3}{2}x - 2$ |
| 150. | a. $y = 5$  b. $x = -2$ |
| 151. | $(-2, 5)$ |

57

# HOMEWORK & EXTRA PRACTICE SCENARIOS

As you complete scenarios in this part of the book, you will practice what you learned in the guided discovery sections. You will develop a greater proficiency with the vocabulary, symbols and concepts presented in this book. Practice will improve your ability to retain these ideas and skills over longer periods of time.

There is an Answer Key at the end of this part of the book. Check the Answer Key after every scenario to ensure that you are accurately practicing what you have learned. If you struggle to complete any scenarios, try to find someone who can guide you through them.

# CONTENTS

Section 1
# REVIEW MULTIPLYING POLYNOMIALS

1. The following multiplication scenarios are typically challenging.  Work through each one as quickly as you can.  Accuracy is more important than speed, though.

    a.  9×7        b.  7×4        c.  8×9        d.  7×8        e.  6×9        f.  7×6

2. Write down the entire prime factorization for each number shown.

    a.  32                                b.  36

3. Write the prime factorization for each number.

    a.  48                      b.  56                            c.  63

4. The number 20 has the following <u>positive</u> factor pairs.  Note: factor pairs must <u>only be</u> integers.

    1×___ , 2×___ , 3×___ , and 4×___

5. Write down the <u>positive</u> factor pairs for each number shown below.

    a.  32                                b.  36

6. Write down the <u>positive</u> factor pairs for each number.

    a.  48                      b.  56                            c.  63

7. Multiply each expression.

    a. $2x(x-1)$          b. $-5x(3x-2)$          c. $-x(3x^2-2x+1)$

8. Rewrite each expression as the product of a monomial and a binomial.

    a. $4B^2+24B$              b. $32x^2-24x$

9. In the previous scenario, how can you be sure that your product is a correctly rewritten form of the original expression?

10. Draw a rectangle that has the following area. Label the length of the base and height.

    a. $13x+26$               b. $20x^2+5x+10$

11. Multiply each expression.

    a. $(x+7)(x+2)$          b. $(x+6y)(x-6y)$

12. Multiply each expression.

    a. $(2x-5)(x+11)$        b. $(x+4)^2$

13. Simplify the expression as much as you can.

$$(x+3)(x^2-3x+9)$$

Section 2
# FINDING A GREATEST COMMON FACTOR

As a reminder, when you rewrite an expression as the product of two other expressions, the two expressions are called factors. In this sense, the word *factor* can be a <u>noun</u>. Furthermore, the process of rewriting an expression as a product of its factors is often referred to as factoring. In this sense, then, the word *factor* can be a <u>verb</u>.

14. Now that we have defined some new vocabulary, factor the following polynomials.

   a.  $3x - 15$
   
   b.  $14x^2 + 6x - 10$

15. Factor each polynomial.

   a.  $6x^2 + 12x + 27$
   
   b.  $12x^4 - 30x^3 - 24x^2$

Identifying a common factor can lead to different opinions. Consider the polynomial $12x^4 - 30x^3 - 24x^2$. The three terms of this trinomial share 2 as a common factor, as well as 3, 6, $x$, $x^2$, $6x$, and $6x^2$. In scenarios like this, it is typical to identify $6x^2$ as the greatest common factor (GCF).

16. The terms in the expression below have many factors in common. Identify all of the common factors of the terms in each expression.

   a.  $24x^3 - 12x^2 + 20$
   
   b.  $6x^2 + 12x^3$

17. Identify the <u>greatest</u> common factor of each expression in the previous scenario.

18. Factor the trinomial.

   $30x^6 - 60x^3 - 45x^5$

19. Factor the trinomial.

   $20x^3 + 30x^4 - 28x^2$

*Section 3*
# WRITING A TRINOMIAL AS A PRODUCT OF TWO BINOMIALS

20. Multiply the following binomials. Notice how the terms in the result are related to the terms in the original binomials.

    a. $(x+5)(x+6)$           b. $(x-5)(x+5)$          c. $(x-4)^2$

In an earlier scenario, the factored form of polynomial was the product of a monomial and a trinomial. In the previous scenario, the factored form is the product of two binomials. Some polynomials cannot be factored at all, such as $x^2+2x+5$.

21. Look at your work for the 3 multiplication scenarios in the previous scenario. Could you do that work in reverse? In other words, if you started with your final expression, could you work backwards to figure out the two binomials that you multiplied to produce that expression? Try to do that with the following expressions. Write each expression as the product of two binomials.

    a. $x^2+5x+6$                b. $x^2+7x+12$

22. Write each expression as the product of two binomials.

    a. $x^2+10x+16$            b. $x^2-x-6$

23. Write each expression as the product of two binomials.

    a. $x^2-4x-12$             b. $x^2+15x-16$

24. What value of $T$ will make the two expressions equivalent?

    a. $x^2+Tx-20$ and $(x-2)(x+10)$

    b. $x^2+Tx-64$ and $(x+8)(x-8)$

    c. $x^2+Txy+16y^2$ and $(x+4y)^2$

25. Once again, multiply the following binomials and try to make connections between the terms in your final expression and the original binomials that you multiplied.

    a. $(x+9)(x-1)$          b. $(x-9y)(x+9y)$          c. $(x-7y)^2$

26. Once again, try to work backwards to write each expression as the product of two binomials.

    a. $x^2+12x+11$          b. $x^2-36$          c. $x^2+20xy+100y^2$

The single most important mental activity involved in factoring is the process of multiplying your factors together to check your answer. You will also find that it helps to start out by *trying* to factor out a GCF from the entire expression. You may not be *able* to factor out a GCF, but if you can, it may help you find additional factors more easily.

27. Fill in the blank to make each expression equal to the factors written below the expression.

    a. $x^2+\underline{\quad}x+18$          b. $x^2+\underline{\quad}x-25$          c. $x^2-7x+12$

       $(x-9)(x-2)$              $(x+5)(x-5)$            $(x+\underline{\quad})(x-4)$

28. Multiply the expressions below to create an equivalent expression without parentheses.

    $10(x+2)(x+5)$

29. When you try to multiply the expressions below, it may be confusing to know how to proceed with the "10" to the left of the first binomial. You may find it easier to first multiply the binomials before distributing the "10", but you can also start by distributing the "10" to the first binomial. Try a third option. Distribute the "10" to both of the binomials and then multiply the binomials. Write down your simplified result to see if it matches the other two options.

30. Multiply the expressions below to find an equivalent expression without parentheses.

$$-4(x+3)(x-2)$$

31. Fill in the blank to make each expression equal to the factors written below the expression.

a.  $2x^2 + \underline{\quad} x - 80$
    $2(x-8)(x+5)$

b.  $3x^2 + \underline{\quad} x - 24$
    $3(x+4)(x-2)$

c.  $-x^2 + 18x - 32$
    $-(x-2)(x+\underline{\quad})$

32. For the trinomial $-x^2 + 14x - 24$, which factorization is correct?

$$-(x-8)(x-3) \quad \text{or} \quad -(x-4)(x-6)$$

33. Each trinomial below has been factored but the plus or minus signs are missing. Fill in either a plus sign (+) or a minus sign (−) to make the factors correct.

a.  $x^2 - 33x + 200$
    $(x \quad 25)(x \quad 8)$

b.  $x^2 + 73x - 150$
    $(x \quad 2)(x \quad 75)$

c.  $x^2 + 29x + 100$
    $(x \quad 4)(x \quad 25)$

34. Each trinomial below has been factored but some of the signs and numbers are missing. Fill in any missing details to make the factors correct.

a.  $x^2 + 16x + 39$
    $(x \quad 13)(x \quad 3)$

b.  $x^2 - 8x + 7$
    $(x \quad 1)(x \quad 7)$

c.  $x^2 - 5x - 126$
    $(x \quad 9)(x \quad 14)$

35. Factor the following trinomials.

a.  $x^2 + 7x + 6$

b.  $x^2 - 7x + 6$

36. Factor the following trinomials.

a.  $x^2 - x - 6$

b.  $x^2 + x - 6$

37. Factor each polynomial below by writing it as the product of a monomial and a remaining polynomial. The monomial should be the greatest common factor of the original polynomial.

   a.  $-2x^2 + 6x$

   b.  $-2x^2 - 10x + 14$

   c.  $15x^4 - 35x^3 - 55x^2$

38. Factor the following trinomials. Check your answer by multiplying your factors together. By doing this, you will know if your factors are correct.

   a.  $x^2 - 5x - 6$

   b.  $20x^2 - 100x - 120$

The previous scenario is designed to show you that it helps to start out by *trying* to factor out a GCF from the entire expression. You may not be *able* to factor out a GCF, but if you can, it may help you find additional factors more easily.

39. Factor each trinomial below.

   a.  $2x^6 - 10x^5 - 12x^4$

   b.  $-3x^2 + 15x + 18$

40. Factor each trinomial below.

   a.  $-x^3 + 5x^2 + 6x$

   b.  $3x^2 + 18x - 21$

41. Factor the following trinomials. Verify that your factors are correct.

   a.  $-5x^2 + 40x - 60$

   b.  $2x^2 - 7x - 9$

Factoring the trinomial in part b. in the previous scenario can be challenging. When you begin to learn about factoring, it is easier to find the factors of a trinomial when the leading coefficient is 1 (as in $1x^2$, or simply $x^2$). If the leading coefficient is any number <u>other than 1</u>, you need to pay close attention to that coefficient and check your factors to see if they are correct.

42. Identify the leading coefficient in each trinomial.

    a. $4x^2 - 3x - 7$  b. $2x^2 + 7x + 3$  c. $-x^2 - 3x + 4$

43. It may be helpful to go through the mental cycle of multiplying binomials one more time. Multiply the following binomials and try to make connections between the terms in your final expression and the original binomials that you multiplied.

    a. $(2x-5)(x+2)$  b. $(3x+7)(x-1)$  c. $(3x-1)(2x+3)$

44. Fill in the blank to make each expression equal to the factors written below the expression.

    a. $2x^2 + \underline{\quad} x - 7$
    $(2x+7)(x-1)$

    b. $3x^2 + \underline{\quad} x - 10$
    $(3x+2)(x-5)$

    c. $4x^2 + \underline{\quad} x - 7$
    $(2x-1)(2x+7)$

45. Fill in the blank to make each expression equal to the factors written below the expression.

    a. $3x^2 - 5x - 2$
    $(\underline{\quad}+1)(\underline{\quad}-2)$

    b. $5x^2 - 13x - 6$
    $(\underline{\quad}-3)(\underline{\quad}+2)$

    c. $6x^2 - 7x - 3$
    $(\underline{\quad}+1)(\underline{\quad}-3)$

46. What value of $U$ will make the two expressions equivalent?

    a. $3x^2 + Ux - 24$ and $(3x-8)(x+3)$

    b. $5x^2 + Ux - 15$ and $(x+5)(5x-3)$

    c. $12x^2 + Ux - 3$ and $(6x+1)(2x-3)$

47. Rewrite each trinomial in Standard Form and identify the leading coefficient in that trinomial.

    a. $1+5x^2-6x$              b. $9x+7+2x^2$              c. $3-2x-x^2$

48. Factor the previous three trinomials. Check your result by multiplying your factors together.

49. Factor the following trinomials. Check your result by multiplying your factors together.

    a. $4x^2-7-3x$              b. $3+2x^2+7x$              c. $-3x+4-x^2$

50. Factor the following trinomials. Check your result by multiplying your factors together.

    a. $5x^2-6+7x$              b. $6x^2-17x+10$           c. $-x^2+12x-36$

At this point, there isn't much more to learn about factoring. In order to strengthen your long-term understanding of this topic, though, we will continue to look more closely at factoring and try to sort out some of the patterns that may help you factor expressions more quickly.

51. Factor the following trinomials.

    a. $15-11x+2x^2$                       b. $30-22x+4x^2$

Once again, the previous scenario is designed to show you that it helps to start out by *trying* to factor out a GCF from the entire expression.

              72              

52. Factor each trinomial below.

a. $-15x + 11x^2 - 2x^3$

b. $20x^2 - 110x + 150$

53. Factor the trinomial.

$-6x^2 + 33x - 45$

54. Explain why $(2x - 1)(2x - 6)$ cannot be the factorization of $4x^2 - 12x + 6$.

55. Factor each trinomial.

a. $x^2 - 49$

b. $4x^2 - 25y^2$

Section 4

# FACTORING A DIFFERENCE OF TWO SQUARES

56. Factor the following expressions. Check your answer by multiplying your factors together.

    a. $x^2+0x-36$                b. $x^2+0x-49$              c. $64+0x-x^2$

57. Factor the following expressions. Check your answer by multiplying your factors together.

    a. $81+0x-x^2$              b. $x^2-100$                 c. $121-x^2$

58. Factor the following expressions. Why are they more difficult than the ones in the previous scenario?

    a. $-3x^2+27$                     b. $-x^4+x^2$

59. Factor the following expressions. Check your answer by multiplying your factors together.

    a. $6x^2-24$                       b. $24-6x^2$

60. Factor the following expressions. Check your answer by multiplying your factors together.

    a. $-5x^2+45$                    b. $810-10x^2$

61. Factor each expression. Check your answer by multiplying your factors together.

    a. $64x-x^3$                      b. $-2x^3+98x$

62. Factor the following expressions.

   a. $36x^2 - 121$

   b. $x^2 - 144y^2$

Many of the previous polynomials (sometimes after factoring out a GCF) have a form of $A^2 - B^2$. This form is known as the difference of two squares. When you identify this form, it will always factor quickly into the following two binomials: $(A + B)(A - B)$.

63. Factor the expression $2x^2 - 32$.

64. Factor the expression $x^4 - 16$.

65. Factor the expression $16x^4 - 1$.

66. If $x^2 - y^2$ can be factored as $(x+y)(x-y)$, how can $x^2 + y^2$ be factored?

*Section 5*
# FACTORING A PERFECT SQUARE TRINOMIAL

67. Factor the following trinomials. Check your answer by multiplying your factors together.

      a. $x^2 - 2x + 1$                              b. $x^2 - 12x + 36$

68. Factor the following trinomials. Check your answer by multiplying your factors together.

      a. $x^2 + 14x + 49$                       b. $4x^2 + 4x + 1$

69. Factor the following trinomials. Check your answer by multiplying your factors together.

      a. $25x^2 + 70x + 49$                   b. $x^2 + x + \dfrac{1}{4}$

70. Factor the following trinomials. Check your answer by multiplying your factors together.

      a. $x^2 + \dfrac{2}{3}x + \dfrac{1}{9}$                b. $x^2 + \dfrac{4}{5}x + \dfrac{4}{25}$

71. Factor the following trinomials.

      a. $x^2 + \dfrac{7}{5}x + \dfrac{49}{100}$               b. $12x - 3x^2 - 12$

72. Factor the following trinomials.

      a. $64x + x^3 - 16x^2$                    b. $4xy - 4y^2 - x^2$

When you factor the previous groups of trinomials, they each involve a trinomial that has a repeated factor. Just as $49P^2$ is a perfect square because it can be written as $\left(7P\right)^2$, these trinomials are called perfect square trinomials.

73. One type of perfect square trinomial will always factor into $\left(F+G\right)\left(F+G\right)$ or $\left(F+G\right)^2$. Multiply these factors back together to see the original trinomial.

74. A perfect square trinomial may also sometimes factor into $\left(C-D\right)\left(C-D\right)$ or $\left(C-D\right)^2$. Multiply these factors back together to see the original trinomial.

75. Factor the expression $M^2-16MN+64N^2$.

76. Explain why $\left(4x-7\right)\left(x-5\right)$ cannot be the factorization of $4x^2-20x+35$.

77. Factor the following trinomials.

    a. $x^2+2xy+y^2$          b. $4x^2-12xy+9y^2$          c. $25x^2-100x+100$

78. Explain how to determine if a trinomial is a perfect square trinomial.

*Section 6*
# USING FACTORING TO SOLVE EQUATIONS

Consider some scenarios that involve polynomials like the ones you have been working with so far.

79. A diver stands at the top of a cliff looking down on a deep lake below. When the diver jumps, her path through the air is modeled by the equation $H = -16t^2 + 64$, where $H$ is the height of the diver, in feet, $t$ seconds after she jumps. What is the height of the diver at the exact moment she jumps?

80. In the previous scenario, how long is the diver in the air before she hits the water?

When the diver hits the water, her height is 0. In the equation $H = -16t^2 + 64$, if you replace $H$ with 0, the equation becomes $0 = -16t^2 + 64$. You can try to isolate "$t$" in this equation, but you may not know how to solve this type of equation yet.

For now, consider simpler equations that show an expression equal to 0. As you become familiar with solving more complex forms of this type of equation, you may be able to come back and answer the question in the previous scenario.

81. What x-value makes each equation true?

    a. $0 = x + 2$            b. $x - 3 = 0$            c. $0 = x + 4$

82. What x-value makes each equation true?

    a. $5(x-3)=0$          b. $-2(x+1)=0$          c. $-3(x-7)=0$

83. What values of x will make each equation true?

    a. $(x-6)(x+1)=0$      b. $(x-10)(x-9)=0$      c. $(x+8)(x+11)=0$

84. In the scenarios above, the factors have a product of 0. Why does this make it possible to determine values of x that will make each equation true?

             

85. Solve the following equations.

    a. $x^2+7x+10=0$          b. $x^2-8x+15=0$          c. $x^2+x-20=0$

86. The equation $(F)(G)=0$ is true if $F = 0$. It is also true if $G = 0$. Suppose, instead that $(F)(G)=2$. What values of $F$ and $G$ make this statement true?

87. Solve the equation $(x+3)(x+4)=3$ by setting both factors equal to 3. Check your solutions to see if either one makes the original equation true.

88. Which equation is easier to solve? Why is this?

    Equation 1: $(x+3)(x+4)=2$          Equation 2: $(x+3)(x+4)=0$

89. It is difficult to find values of $x$ that make Equation 1 true in the previous scenario. It will become easier to solve, though, if you can find a way to change the equation. Try to change Equation 1 to make it look more like Equation 2 in the previous scenario.

90. Solve the equation $(x-2)(x+6)=9$.

91. What $x$-value makes each equation true?

    a. $5x+3=0$          b. $7x-1=0$

92. Solve each equation.

    a. $2(5x+3)=0$                            b. $10(7x-1)=0$

93. What values of $x$ will make each equation true?

    a. $(4x-1)(x-5)=0$       b. $(x+8)(8x-1)=0$       c. $(3x+2)(4x-5)=0$

94. Solve the following equations.

    a. $10=6x^2+28x$         b. $20x^2-45=0$         c. $9x^2+4=12x$

95. What values of $x$ will make each equation true?

    a. $x(x-2)=0$              b. $x(x+3)=0$              c. $5x(2x+3)=0$

96. In each of the previous three equations, one of the solutions is 0. Why does this occur?

97. Solve the following equations.

    a. $x^2+2x=0$             b. $x^2-3x+5=5$        c. $3=10x^2-15x+3$

98. What values of $x$ will make each equation true?

        a. $x^3(x-11)=0$                                 b. $7x^2(x+4)=0$                         ★c. $2x(x^2-1)=0$

99. Solve the following equations in two different ways. For the first approach, factor as you have learned to do in the previous scenarios. For the second approach, divide both sides by $x$ and then proceed with factoring. What do you notice?

        a. $x^2=10x$                                       b. $x^3-2x^2+x=0$

100. Solve the following equations.

        a. $4x^2-21x+5=0$                  b. $8x^2+63x-8=0$            c. $12x^2-7x-10=0$

101. In a previous scenario, you saw the equation below, but you may not have known how to solve it. Try to solve it now.

        $-16t^2+64=0$

102. Solve the equation $(x-6)(x-1)=14$. Check your solutions.

103. Solve the equation $\frac{1}{6}x^2-\frac{1}{3}x-\frac{1}{2}=0$.

104. The fractions in the previous equation make it challenging. Consider another equation with fractions:
$\frac{1}{8}x=\frac{1}{2}$. How can you alter this equation to make it easier to solve?

105. Solve the equation by multiplying both sides by a value that will eliminate the fractions.

$$\frac{1}{6}x-\frac{2}{3}=\frac{1}{2}$$

106. Solve each equation by multiplying both sides by a value that will eliminate the fractions. Try to find the smallest value that will do this, to keep the numbers as small possible.

a. $\frac{1}{4}x^2-\frac{5}{4}x+1=0$      ★b. $x^2-7x+8.25=0$

107. Fill in the blank. To solve the equation below, it would be helpful to first multiply both sides by the number ____ to clear all of the fractions on both sides. Do <u>not</u> solve the equation.

$$\frac{2x}{5}-1=\frac{1}{3}x-\frac{5}{6}$$

108. Finish solving the previous equation.

109. Solve the equation shown.

$$x^3-2x^2=15x$$

Section 7
# SCENARIOS THAT INVOLVE FACTORING

110. A kid builds a small catapult and uses it to launch a rock. The path of the rock is modeled by the equation $H=-x^2+5x$, where $H$ is the height of the rock after it is launched, in feet, and $x$ is the distance the rock has moved horizontally, in feet. How far away from the kid does the rock land when it hits the ground?

111. A bug sits on a leaf of a plant and gets ready to jump down into a puddle on the ground. A frog waits patiently on the ground. When the bug jumps, its path is modeled by the equation $H=-16t^2+2t+4$. In the equation, $H$ is the height of the frog in inches and $t$ is the time that has passed after the bug jumps, measured in seconds. If the frog wants to catch the bug with its tongue when the bug is 1 inch above the ground, how long will the frog need to wait after the bug jumps to extend its tongue? Assume the frog catches the bug at the exact moment that it reaches out its tongue.

112. The area of the rectangle shown is 12 in².

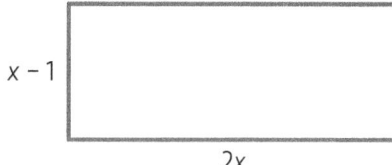

x − 1

2x

   a. What is the value of $x$?

   b. What is the perimeter of the rectangle?

113. The length of a rectangle is 3 more than twice its width. The area of the rectangle is 44 square centimeters. What is the perimeter of the rectangle?

114. Rabbits and deer like to eat the plants in your garden, so you buy enough fencing to surround your rectangular garden. The area of the rectangle that you can surround with fencing is modeled by the equation $A = w(40-w)$, where $A$ is the area of the rectangle if it has a width of $w$ feet.

    a. What is the area of the rectangle, if the width is 10 feet?

    b. How many feet of fencing do you need to buy if the width is 10 feet?

115. ★In the previous scenario, what is the width of the rectangle if the area is 357.75 square feet?

116. What is the perimeter of the outer rectangle if the shaded region has an area of 92 square units?

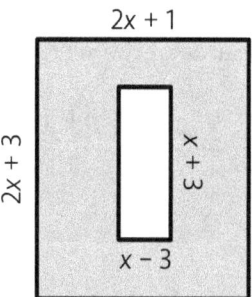

117. The area of a triangle is 20 square inches. Its height is 2 inches longer than twice the length of its base. What are the dimensions of the triangle?

118. Equations like 3x − 5 = 8 and −2x + 3 = x usually have one solution and they are known as <u>linear</u> equations.  Solve the following linear equations.

    a.  $3(x-6)=5x-4$                                  b.  $1-3(2x-1)=19$

119. Equations like $x^2$ + 3x + 2 = 0 and $x^2$ + 5x = 6 are called <u>quadratic</u> equations and they can be rearranged to look like $Ax^2 + Bx + C = 0$.  Solve the following equations.

    a.  $(y-7)(y+2)=0$                                  b.  $3y^2 - 7y = 20$

120. Solve the equation $x^2 = 25$.

121. ★The two solutions to a quadratic equation have a sum of −7 and a product of 12.  The solutions are integers.  What is the equation?  Write the equation in the form $Ax^2 + Bx + C = 0$.

## Section 8
# USING FACTORING TO SIMPLIFY FRACTIONS

122. Simplify each of the following fractions.

a. $\dfrac{2\cdot 10}{3\cdot 10}$

b. $\dfrac{2\cdot 3}{3\cdot 5}$

c. $\dfrac{3x}{8x}$

d. $\dfrac{8y}{12y}$

123. Simplify each of the following fractions.

a. $\dfrac{3x(x+4)}{5x(x+4)}$

b. $\dfrac{(x+1)(x+3)}{(x+1)(x+7)}$

c. $\dfrac{6x+6}{7x+7}$

124. In the previous scenario, the last fraction can only be simplified after you factor the expressions in the numerator and denominator. After you factor, it becomes clear that there are identical factors in the numerator and denominator. Use this strategy to simplify each of the following fractions.

$$\dfrac{x^2+4x+3}{x^2+8x+7}$$

125. Simplify each fraction by cancelling out disguised forms of 1.

a. $\dfrac{x^2-1}{x^2+2x+1}$

b. $\dfrac{x^2+6x+5}{x^2-25}$

126. Simplify the following fraction: $\dfrac{6x^2-27x-15}{9x^2-42x-15}$ .

Section 9
# INTRODUCTION TO GRAPHING PARABOLAS

127. If you graph every ordered pair for the equation $y = \frac{1}{2}x - 2$ what will be the shape of your graph? Draw the graph.

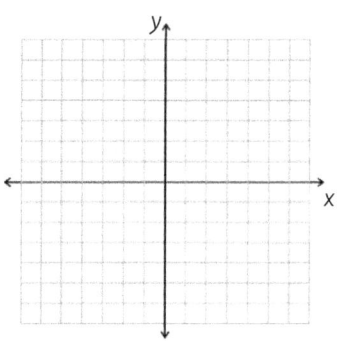

128. When you graph all of the ordered pairs represented by the equation $y = x^2 - 2$, the shape of the graph is. . . actually, let's see if you can figure out the shape of the graph. Use what you know about graphing equations to create a thorough graph of this equation.

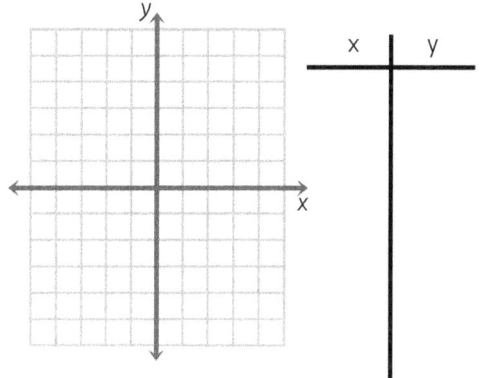

The graph in the previous scenario has a curved shape that makes it very different than linear graphs. In mathematics, there is a technical name for practically everything. This specially curved graph is called a parabola (puh-RA-buh-luh). Additionally, while the structure $y = \frac{1}{2}x - 2$ is called a linear equation, the structure $y = x^2 - 2$ is called a quadratic equation.

129. Graph the equation $y = -x^2 + 1$.

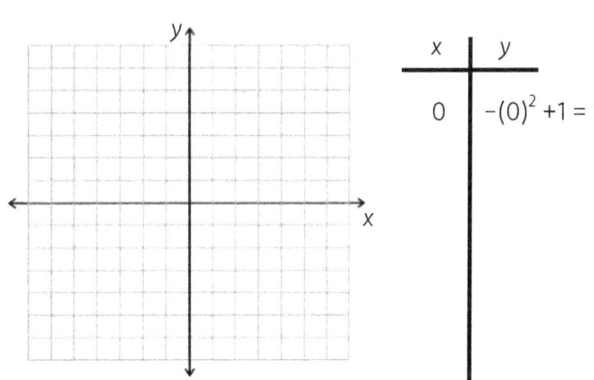

| x | y |
|---|---|
| 0 | $-(0)^2 + 1 = 1$ |

130. Consider the parabola to the right.

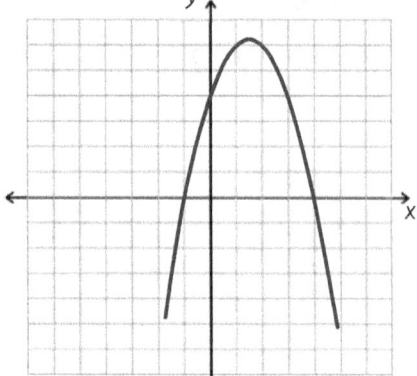

       a.  Determine the x-value(s) for which the parabola has a y-value of 0.

       b.  Determine the y-value(s) for which the parabola has an x-value of 3.

       c.  Determine the x-value(s) for which the parabola has a y-value of 6.

131. Determine the x-value(s) for which the equation $y = -x^2 + 3x + 10$ has a y-value of

       a.  0                                   ★b.  10

132. Identify any x- and y-intercepts for each of the following equations.

       a.  $3x + 2y = 12$                        b.  $y = x^2 - 2x - 8$

133. ★Identify any x- and y-intercepts for each of the following equations.

       a.  $x^2 + y = 3$                         b.  $y = 4x^2 - 12x + 9$

134. ★Draw a <u>very basic</u> sketch of a parabola that has each of the following characteristics.

a. <u>no x-intercepts</u>

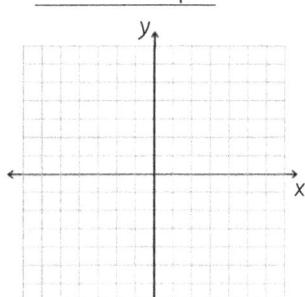

b. <u>one x-intercept</u>

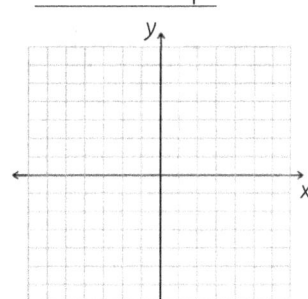

c. <u>two x-intercepts</u>

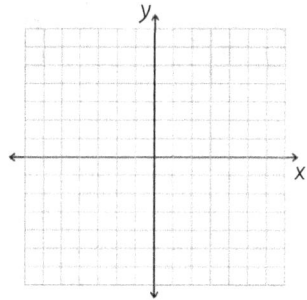

135. ★Identify the x- and y-intercepts of the following equation. Then, graph the equation to verify the accuracy of your intercepts.

$$y = x^2 - 2x - 3$$

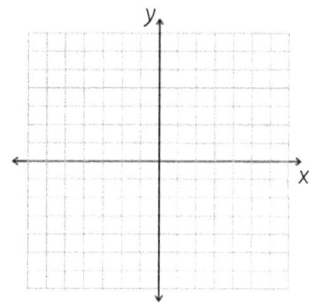

136. Part of a parabola is shown below. Even though you do not have the equation for the parabola, use what you have learned so far to plot more points that are on the parabola. Use those points to finish drawing the parabola.

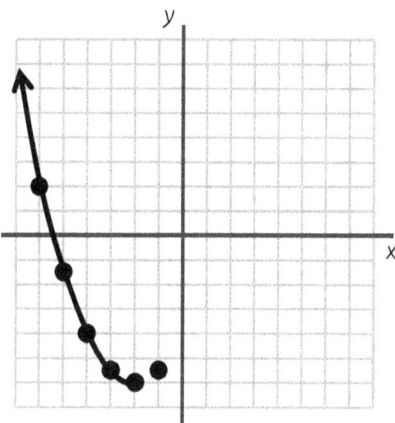

# Section 10
# CUMULATIVE REVIEW

137. Simplify each expression by eliminating parentheses and using only positive exponents.

a. $\dfrac{x^{-2}}{x^{-3}}$

b. $\left(-\dfrac{1}{3}\right)^{-3}$

c. $2\left(x^{-1}\right)^{2}$

138. The average (arithmetic mean) of 7, 13, 21, and $H$ is 10. Determine the value of $H$.

139. Write the equation for both lines shown, in Slope-Intercept Form.

a. Line 1

b. Line 2

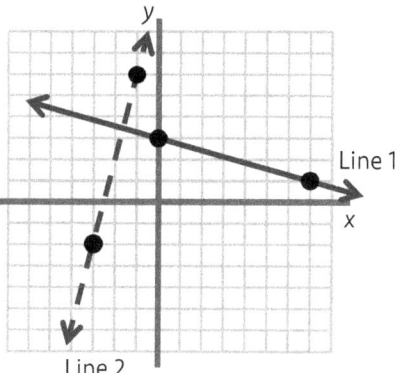

c. Are the two lines perpendicular? How do you know?

140. The average TV viewership for the Super Bowl from 2006 to 2015 (a 10-year period) was 104.3 million people. When you include the 2016 Super Bowl, the average viewership from 2006 to 2016 was 105 million people. How many people watched the 2016 Super Bowl?

141. Solve each equation.

   a. $7x-6=\dfrac{1}{3}\left(6x-3\right)$

   b. $-x+9=-\dfrac{1}{5}x+2$

142. If the slope of the line $y=2+7x$ is changed to become $-4$, what is the new equation of the line?

143. What is the slope of the line $3x=2\left(y-6\right)$?

144. A parenting survey was conducted in 2015 in a local town to find out what challenges parents face as they raise their children. 60% of fathers in the town chose to take the survey, while 40% of mothers chose to take the survey.

   a. If 55% of the parents in the town are mothers, what percent of the parents in the entire town chose to take the survey?

   b. A total of 6,125 parents took the survey in 2015. What was the total number of parents in the town in 2015?

145. Circle the equations that are quadratic.

   a. $2x^2-x=6$

   b. $2x-6=x$

   c. $3x=4-x^2$

## Section 11
# ANSWER KEY

| | |
|---|---|
| 1. | a. 63    b. 28    c. 72    d. 56    e. 54    f. 42 |
| 2. | a. $2^5$              b. $2^2 \times 3^2$ |
| 3. | a. $2^4 \times 3$      b. $2^3 \times 7$      c. $3^2 \times 7$ |
| 4. | 20, 10, none, 5 |
| 5. | a. $1 \times 32, 2 \times 16, 4 \times 8$<br>b. $1 \times 36, 2 \times 18, 3 \times 12, 4 \times 9, 6 \times 6$ |
| 6. | a. $1 \times 48, 2 \times 24, 3 \times 16, 4 \times 12, 6 \times 8$<br>b. $1 \times 56, 2 \times 28, 4 \times 14, 7 \times 8$<br>c. $1 \times 63, 3 \times 21, 7 \times 9$ |
| 7. | a. $2x^2 - 2x$        b. $-15x^2 + 10x$<br>c. $-3x^3 + 2x^2 - x$ |
| 8. | a. $4B(B+6)$          b. $8x(4x-3)$ |
| 9. | Multiply the monomial and binomial to see if you get the original expression. |
| 10. | $\boxed{13 \atop x+2}$   $\boxed{5 \atop 4x^2+x+2}$ |
| 11. | a. $x^2 + 2x + 7x + 14 \rightarrow x^2 + 9x + 14$<br>b. $x^2 - 6xy + 6xy - 36y^2 \rightarrow x^2 - 36y^2$ |
| 12. | a. $2x^2 + 22x - 5x - 55 \rightarrow 2x^2 + 17x - 55$<br>c. $(x+4)(x+4) \rightarrow x^2 + 8x + 16$ |
| 13. | $x^3 - 3x^2 + 9x + 3x^2 - 9x + 27 \rightarrow x^3 + 27$ |
| 14. | a. $3(x-5)$      b. $2(7x^2 + 3x - 5)$ |
| 15. | a. $3(2x^2 + 4x + 9)$      b. $6x^2(2x^2 - 5x - 4)$ |
| 16. | a. 2, 4    b. 2, 3, 6, 2x, 3x, 6x, $2x^2$, $3x^2$, $6x^2$ |
| 17. | a. 4              b. $6x^2$ |
| 18. | $15x^3(2x^3 - 4 - 3x^2)$ |
| 19. | $2x^2(10x + 15x^2 - 14)$ |
| 20. | a. $x^2 + 11x + 30$    b. $x^2 - 25$    c. $x^2 - 8x + 16$ |
| 21. | a. $(x+2)(x+3)$        b. $(x+3)(x+4)$ |
| 22. | a. $(x+8)(x+2)$        b. $(x-3)(x+2)$ |
| 23. | a. $(x-6)(x+2)$        b. $(x+16)(x-1)$ |
| 24. | a. $T = 8$      b. $T = 0$      c. $T = 8$ |

| | |
|---|---|
| 25. | a. $x^2 + 8x - 9$        b. $x^2 - 81y^2$<br>c. $x^2 - 14xy + 49y^2$ |
| 26. | a. $(x+11)(x+1)$      b. $(x+6)(x-6)$<br>c. $(x+10y)(x+10y)$ or $(x+10y)^2$ |
| 27. | a. −11        b. 0        c. −3 |
| 28. | $10(x^2 + 7x + 10) \rightarrow 10x^2 + 70x + 100$ |
| 29. | $(10x+20)(10x+50) \rightarrow 100x^2 + 700x + 1000$<br>When you distribute the 10 to both binomials, the result is 10 times larger than it should be. |
| 30. | $-4(x^2 + x - 6) \rightarrow -4x^2 - 4x + 24$ |
| 31. | a. −6        b. 6        c. −16 |
| 32. | Neither. It is $-(x-12)(x-2)$. |
| 33. | a. $(x-25)(x-8)$      b. $(x-2)(x+75)$<br>c. $(x+4)(x+25)$ |
| 34. | a. $(x+13)(x+3)$      b. $(x-1)(x-7)$<br>c. $(x+9)(x-14)$ |
| 35. | a. $(x+6)(x+1)$        b. $(x-1)(x-6)$ |
| 36. | a. $(x-3)(x+2)$        b. $(x+3)(x-2)$ |
| 37. | a. $-2x(x-3)$ or $2x(-x+3)$<br>b. $-2(x^2 + 5x - 7)$ or $2(-x^2 - 5x + 7)$<br>c. $5x^2(3x^2 - 7x - 11)$ |
| 38. | a. $(x-6)(x+1)$        b. $20(x-6)(x+1)$ |
| 39. | a. $2x^4(x-6)(x+1)$      b. $-3(x-6)(x+1)$ |
| •40. | a. $-x(x-6)(x+1)$      b. $3(x+7)(x-1)$ |
| 41. | a. $-5(x-6)(x-2)$<br>b. $(2x-9)(x+1)$ |
| 42. | a. 4              b. 2              c. −1 |
| 43. | a. $2x^2 - x - 10$      b. $3x^2 + 4x - 7$<br>c. $6x^2 + 7x - 3$ |

99

| | |
|---|---|
| 44. | a. 5  b. -13  c. 12 |
| 45. | a. $(3x+1)(x-2)$  b. $(x-3)(5x+2)$  c. $(3x+1)(2x-3)$ |
| 46. | a. 1  b. 22  c. -16 |
| 47. | a. $5x^2-6x+1$  b. $2x^2+9x+7$  c. $-x^2-2x+3$ |
| 48. | a. $(5x-1)(x-1)$  b. $(2x+7)(x+1)$  c. $(-x-3)(x-1)$ or $-(x+3)(x-1)$ |
| 49. | a. $(4x-7)(x+1)$  b. $(2x+1)(x+3)$  c. $-(x+4)(x-1)$ or $(-x-4)(x-1)$ or $(x+4)(-x+1)$ |
| 50. | a. $(5x-3)(x+2)$  b. $(6x-5)(x-2)$  c. $-(x-6)(x-6)$ or $(-x+6)(x-6)$ |
| 51. | a. $(2x-5)(x-3)$  b. $2(2x-5)(x-3)$ |
| 52. | a. $-x(2x-5)(x-3)$  b. $10(2x-5)(x-3)$ |
| 53. | $-3(2x-5)(x-3)$ |
| 54. | $(2x-1)(2x-6) = 4x^2-14x+6$ |
| 55. | a. $(x+7)(x-7)$  b. $(2x+5y)(2x-5y)$ |
| 56. | a. $(x+6)(x-6)$  b. $(x+7)(x-7)$  c. $(8+x)(8-x)$ |
| 57. | a. $(9+x)(9-x)$  b. $(x+10)(x-10)$  c. $(11+x)(11-x)$ |
| 58. | These are more difficult because they have a greatest common factor that can be factored out first.  a. $-3(x+3)(x-3)$  b. $-x^2(x+1)(x-1)$ |
| 59. | a. $6(x+2)(x-2)$  b. $6(2+x)(2-x)$ |
| 60. | a. $-5(x+3)(x-3)$  b. $10(9+x)(9-x)$ |
| 61. | a. $x(8+x)(8-x)$  b. $-2x(x+7)(x-7)$ |
| 62. | a. $(6x+11)(6x-11)$  b. $(x+12y)(x-12y)$ |
| 63. | $2(x+4)(x-4)$ |
| 64. | $(x^2+4)(x^2-4) \rightarrow (x^2+4)(x+2)(x-2)$ |
| 65. | $(4x^2+1)(2x+1)(2x-1)$ |
| 66. | It cannot be factored |
| 67. | a. $(x-1)^2$  b. $(x-6)^2$ |
| 68. | a. $(x+7)^2$  b. $(2x+1)^2$ |

| | |
|---|---|
| 69. | a. $(5x+7)^2$  b. $\left(x+\dfrac{1}{2}\right)^2$ |
| 70. | a. $\left(x+\dfrac{1}{3}\right)^2$  b. $\left(x+\dfrac{2}{5}\right)^2$  c. |
| 71. | a. $\left(x+\dfrac{7}{10}\right)^2$  b. $-3(x-2)^2$ |
| 72. | a. $x(x-8)^2$  b. $-(x-2y)^2$ |
| 73. | $F^2+2FG+G^2$ |
| 74. | $C^2-2CD+D^2$ |
| 75. | $(M-8N)^2$ |
| 76. | $(4x-7)(x-5)=4x^2-27x+35$ |
| 77. | a. $(x+y)^2$  b. $(2x-3y)^2$  c. $25(x-2)^2$ |
| 78. | In a perfect square trinomial, the middle term is twice the product of the square roots of the first and third terms.  Stated another way, find the product of the first and third terms.  Take the square root of that result.  If the middle term is twice the value of that square root, the trinomial is a perfect square trinomial. |
| 79. | 64 feet (Let $t$ = 0.) |
| 80. | 2 sec; Let $H$ = 0.  Solve $0=-16t^2+64$. |
| 81. | a. $x = -2$  b. $x = 3$  c. $x = -4$ |
| 82. | a. $x = 3$  b. $x = -1$  c. $x = 7$ |
| 83. | a. $x = 6, -1$  b. $x = 10, 9$  c. $x = -8, -11$ |
| 84. | If the product of two factors is 0, one of the factors must be 0. |
| 85. | a. $(x+2)(x+5)=0 \rightarrow x=-2, -5$  b. $(x-3)(x-5)=0 \rightarrow x=3, 5$  c. $(x+5)(x-4)=0 \rightarrow x=4, -5$ |
| 86. | Infinitely many options; $F=2, G=1$ or $F=4, G=\dfrac{1}{2}$ or $F=6, G=\dfrac{1}{3}$ etc... |
| 87. | $x = 0$ or $-1$ but neither one makes the original equation true. |
| 88. | #2; The product equals "0," which forces one of the factors to be "0." |
| 89. | $x = -2$ or $-5$  $x^2+7x+12=2 \rightarrow x^2+7x+10=0$ |
| 90. | $x = -7, 3$  $x^2+4x-12=9 \rightarrow x^2+4x-21=0$ |
| 91. | a. $x=-\dfrac{3}{5}$  b. $x=\dfrac{1}{7}$ |

100

| | |
|---|---|
| 92. | a. $x=-\dfrac{3}{5}$      b. $x=\dfrac{1}{7}$ |
| 93. | a. $x=\dfrac{1}{4},5$   b. $x=-8,\dfrac{1}{8}$   c. $x=-\dfrac{2}{3},\dfrac{5}{4}$ |
| 94. | a. $0=2(3x-1)(x+5)\rightarrow x=\dfrac{1}{3},-5$ <br><br> b. $5(2x+3)(2x-3)=0\rightarrow x=-\dfrac{3}{2},\dfrac{3}{2}$ <br><br> c. $9x^2-12x+4=0\rightarrow(3x-2)^2=0\rightarrow x=\dfrac{2}{3}$ |
| 95. | a. $x=0,2$   b. $x=0,-3$   c. $x=0,-\dfrac{3}{2}$ |
| 96. | The factor multiplied by the expression in parentheses can be replaced with 0 to make the equation true. |
| 97. | a. $x=0,-2$   b. $x=0,3$   c. $x=0,\dfrac{3}{2}$ |
| 98. | a. $x=0,11$   b. $x=0,-4$   c. $x=0,1,-1$ |
| 99. | a. $x=0,10$ vs. $x=10$   b. $x=0,1$ vs. $x=1$ <br> If you divide both sides by $x$, you will not see that $x=0$ is one of the solutions. |
| 100. | a. $(4x-1)(x-5)=0\rightarrow x=\dfrac{1}{4},5$ <br><br> b. $(8x-1)(x+8)=0\rightarrow x=\dfrac{1}{8},-8$ <br><br> c. $(4x-5)(3x+2)=0\rightarrow x=\dfrac{5}{4},-\dfrac{2}{3}$ |
| 101. | a. $-16(t^2-4)=0\rightarrow-16(t+2)(t-2)=0$ <br> $t=2,-2$ |
| 102. | $x^2-7x+6=14\rightarrow x^2-7x-8=0$ <br> $(x+1)(x-8)=0\rightarrow x=8,-1$ |
| 103. | $6\cdot\left(\dfrac{1}{6}x^2-\dfrac{1}{3}x-\dfrac{1}{2}\right)=0\cdot6\rightarrow x^2-2x-3=0$ <br><br> $(x-3)(x+1)=0\rightarrow x=3,-1$ |
| 104. | Multiply both sides by 8 |
| 105. | $6\cdot\left(\dfrac{1}{6}x-\dfrac{2}{3}\right)=\left(\dfrac{1}{2}\right)\cdot6\rightarrow x-4=3\rightarrow x=7$ |
| 106. | a. $x=4,1$     b. $x=5.5,1.5$ |
| 107. | $30\rightarrow$ least common multiple of 3, 5 and 6 |
| 108. | $30\cdot\left(\dfrac{2x}{5}-1\right)=\left(\dfrac{1}{3}x-\dfrac{5}{6}\right)\cdot30$ <br><br> $\rightarrow12x-30=10x-25\rightarrow2x=5\rightarrow x=2.5$ |
| 109. | $x^3-2x^2-15x=0\rightarrow x(x^2-2x-15)=0$ <br><br> $\rightarrow x(x-5)(x+3)=0\rightarrow x=0,5,-3$ |
| 110. | 5 feet <br> $0=-x^2+5x\rightarrow0=-x(x-5)\rightarrow x=0$ or $5$ |
| 111. | One-half of a sec.; solve $1=-16t^2+2t+4$ <br><br> $0=(8t+3)(2t-1)\rightarrow t=-\dfrac{3}{8}$ or $\dfrac{1}{2}$ |
| 112. | a. $x=3$; Area $=b\cdot h$: solve $2x(x-1)=12$ <br> b. P = sum of all sides: P = 16 in. |
| 113. | width: 4cm     length: 11cm <br> Perimeter: $4+4+11+11=30$ cm <br> Solve: $w(2w+3)=44$ <br><br> $2w^2+3w=44\rightarrow2w^2+3w-44=0$ <br><br> $\rightarrow(2w+11)(w-4)=0\rightarrow w=-\dfrac{11}{2}$ or $4$ |
| 114. | a. $300\ \text{ft}^2$      b. 80 feet |
| 115. | $w=13.5$ ft or $26.5$ ft;   solve: <br> $w(40-w)=357.75\rightarrow4w-w^2=357.75$ <br><br> $4\cdot(40w-w^2)=(357.75)\cdot4$ <br><br> $160w-4w^2=1431\rightarrow0=4w^2-160w+1431$ <br> $(2w-27)(2w-53)=0\rightarrow w=13.5$ or $26.5$ ft |
| 116. | perimeter: 40 units <br> solve: $(2x+3)(2x+1)-(x+3)(x-3)=92$ <br> $x=4$ |
| 117. | base = 4 in.; height = 10 in. <br><br> solve $\dfrac{1}{2}(x)(2x+2)=20\rightarrow x^2+x=20$ <br><br> $x^2+x-20=0\rightarrow(x+5)(x-4)$ <br> $x=4$ or $-5$ (length can't be negative) |
| 118. | a. $x=-7$      b. $x=-2.5$ |
| 119. | a. $y=7,-2$      b. $3y^2-7y-20=0$ <br><br> $\rightarrow(3y+5)(y-4)=0\rightarrow y=-\dfrac{5}{3},4$ |
| 120. | 2 solutions: $x=5$ or $-5$ |
| 121. | Equation: $x^2+7x+12=0$ <br> The solutions are $-3$ and $-4$. <br> $x=-3$ or $-4\rightarrow(x+3)(x+4)=0$ <br> $\rightarrow x^2+7x+12=0$ |
| 122. | a. $\dfrac{2}{3}$    b. $\dfrac{2}{5}$    c. $\dfrac{3}{8}$    d. $\dfrac{2}{3}$ |
| 123. | a. $\dfrac{3}{5}$    b. $\dfrac{x+3}{x+7}$    c. $\dfrac{6(x+1)}{7(x+1)}\rightarrow\dfrac{6}{7}$ |
| 124. | $\dfrac{(x+3)(x+1)}{(x+7)(x+1)}\rightarrow\dfrac{x+3}{x+7}$ |

| 125. | a. $\dfrac{(x+1)(x-1)}{(x+1)(x+1)} \rightarrow \dfrac{x-1}{x+1}$ <br><br> b. $\dfrac{(x+5)(x+1)}{(x+5)(x-5)} \rightarrow \dfrac{x+1}{x-5}$ |
|---|---|
| 126. | $\dfrac{3(2x^2-9x-5)}{3(3x^2-14x-5)} \rightarrow \dfrac{3(2x+1)(x-5)}{(3x+1)(x-5)} \rightarrow \dfrac{2x+1}{3x+1}$ |
| 127. | |
| 128. | |
| 129. | |
| 130. | a. $x = -1$ or $4$  b. $y = 4$  c. $x = 1$ or $2$ |
| 131. | a. $x = -2$ or $5$; solve $0 = -x^2+3x+10$ <br> b. $x = 0$ or $3$; solve $10 = -x^2+3x+10$ |
| 132. | To find $x$-intercepts, replace $y$ with 0. <br> To find $y$-intercepts, replace $x$ with 0. <br> a. $x$-int: $(4,0)$; $y$-int: $(0,6)$ <br> b. $x$-ints: $(4,0)$ and $(-2,0)$; $y$-int: $(0,-8)$ |
| 133. | a. $x$-ints: $\left(\sqrt{3},0\right)$ and $\left(-\sqrt{3},0\right)$; $y$-int: $(0,3)$ <br> b. $x$-int: $(1.5,0)$; $y$-int: $(0,9)$ |

| 134. | Draw a parabola that. . . <br> a. does not cross the $x$-axis <br> b. crosses the $x$-axis 1 time <br> c. crosses the $x$-axis 2 time2 |
|---|---|
| 135. | |
| 136. | Plot points at $(0,-4)$, $(1,-1.5)$ and $(2,2)$ <br> |
| 137. | a. $x$   b. $-27$   c. $\dfrac{2}{x^2}$ |
| 138. | $H=-1$; solve $\dfrac{7+13+21+H}{4}=10$ |
| 139. | a. $y=-\dfrac{2}{7}x+3$   b. $y=4x+10$   c. No, the slopes need to be opposite reciprocals to make the lines perpendicular. |
| 140. | 112 million |
| 141. | a. 1   b. 8.75 |
| 142. | $y=2-4x$  or  $y=-4x+2$ |
| 143. | The slope of the line is $\dfrac{3}{2}$. <br><br> $3x=2y-12 \rightarrow 2y=3x+12 \rightarrow y=\dfrac{3}{2}x+6$ |
| 144. | a. Suppose there are 100 in the town <br> 55 mothers $\rightarrow$ .4(55) = 22 took survey <br> 45 fathers $\rightarrow$ .6(45) = 27 took survey <br> 49 took survey <br> 49% of the parents took the survey <br> b. 12,500 parents; solve: $6125 = 0.49x$ |
| 145. | Circle the equations in parts a and c. |

103

www.ingramcontent.com/pod-product-compliance
Lightning Source LLC
Chambersburg PA
CBHW081520220526
45467CB00010B/2984